ALL ABOUT GOATS

ALL ABOUT GOATS

Lois Hetherington
and
John G. Matthews

with a contribution by L. F. Jenner

Farming Press

First edition 1977
Third edition 1992
Reprinted with amendments 1996
Reprinted with amendments 1999

Copyright © Farming Press 1977, 1992

A catalogue record for this book is available
from the British Library

ISBN 0 85236 229 3

**Published by Farming Press,
Miller Freeman UK Ltd,
2 Wharfedale Road,
Ipswich, Suffolk, IP1 4LG,
United Kingdom.**

Distributed in North America by Diamond
Farm Enterprises, Box 537, Bailey
Settlement Road, Alexandria Bay, NY
13607, USA

Colour plates supplied by Hilary Matthews
Typeset by Galleon Photosetting, Ipswich
Printed and bound in Great Britain by
Butler and Tanner Ltd, Frome

CONTENTS

Routine vaccination ● Routine worm control
● Routine treatment for external parasites
● Routine testing for the CAE virus ● Foot care
● The sick goat ● The veterinary first aid box
● Breeding problems ● Control of the breeding cycle
● Problems connected with kidding ● Kids
● Conditions of the udder and teats ● Plant
poisoning ● Skin problems ● Lameness
● Eye infections ● Respiratory diseases
● External lumps

APPENDICES

A colour section appears between pages 20 and 21

ILLUSTRATIONS

PREFACE

THERE is less known about that much maligned animal the goat than the common or garden goldfish, and it is regarded in much the same light—as a pet that does not need to be fed properly.

It will, it is thought, survive staked out on a bare trodden patch of weeds, and moved sometimes . . . if anyone happens to think of it, poor animal. When it is considered that more people in the world use the goat and its products than use cows; that in the Tropics where heat dries up the herbage, goats can eat this hard, seemingly unpalatable material and put it to good use where other animals would not survive, surely it is strange that such a useful creature is so neglected. Only in times of national stress, war and its aftermath, does the goat spring to prominence.

Able to survive in differing climates, from Norway to the Sahara, goats are kept in as many ways as there are people keeping them. Admittedly, residents in high rise flats may find it difficult, but if an allotment is to be had within reasonable distance it might be possible!

Reasons for keeping goats are numerous, and in these days of DIY, with the added demand for health foods, what better than your own fresh dairy foods.

Not many articles are produced especially for goatkeepers, and so the 'Heath Robinsons' in the goat world are in the majority, adapting some of the most improbable objects for their stock—to great advantage. As the animals are of a size which does not require enormous buildings of a special type, all manner of housing is to be found, from chicken huts to garages. A Ministry of Agriculture official, visiting goat herds in the district, remarked to me with understatement, 'I have never met so many great individualists'. Indeed it must be said, goatkeepers do 'come assorted', since goats are to be found in the parks of great castles as well as in garden sheds in urban semidets. One thing all owners have in common is their affection for their animals.

LOIS HETHERINGTON

Saxtead, Suffolk
May, 1976

ACKNOWLEDGMENTS FOR THE THIRD EDITION

I wish to express my thanks to those who have kindly supplied photographs, in particular Mrs P. Carter and Desmond Hetherington; also to my long-suffering husband who prepared the original sketches. I am indebted, too, to John Halliday for providing the bibliography, to John G. Matthews for the diseases and ailments section, to Mr L. F. Jenner for the chapter on commercial goatkeeping and to Robin Pepper MRCVS DBR for the section on fibre and meat.

L.H.

Colchester, Essex
June, 1996

ALL ABOUT GOATS

CHAPTER 1

HOW TO BEGIN

PROBABLY the simplest way to start goatkeeping is to visit the goat section at a big agricultural show, where you will see some good specimens of most breeds and get an idea of what a nice goat should look like. Talk to the exhibitors and maybe you will hear of something for sale. And, of course, you may be able to get the addresses of the owners in your area whom you could contact or visit.

The British Goat Society has been in existence for over 100 years. Affiliated to it are about 110 local goat clubs, so wherever you live in the British Isles, there must be one within reasonable distance. These clubs, though unknown to many people, can be of great help. A list of them is published as Appendix 2.

BUYING STOCK

Sometimes it is possible to buy from advertisements, but caution must be used. If answering them, ask for the registered number and breed of the animal and check the earmark number against the paperwork. If going to view, try to get an experienced goat-keeper to go with you, as so many swans turn out to be geese. Because an animal is in appearance like a recognised breed it will be offered as that, but if there are no papers to prove it, it is in fact an unregistered goat. There are of course some very good unregistered goats which have been sired by good males, but not registered. You might get one, but it is just as likely to be a real scrub. She might produce a fair amount of milk for a short time, then dry up for the autumn and winter, whereas your real dairy goat will continue, with a diminished yield, to milk through the winter, and indeed until the time she is served in the second autumn after kidding—almost two years.

Shy away from dealers; some, I am sure, are all they claim

1

to be, others are decidedly not. Markets are also bad places in which to buy. There may be good reasons for an animal being sold there, but you are more likely to be sold a 'pup'. Do not rush into buying a goat just because it just happens to be there. If you begin with an unsatisfactory animal from stock of unknown origin, the disappointment in store can cause such frustration and loss of interest that you may well give up without any real trial at all.

Having decided to keep goats, contact as many breeders as you can (you can get addresses from the British Goat Society or a local goat club) and see as many animals as possible. There may well not be any of those particular goats you want for sale, but perhaps you could book a kid from one which appealed to you, or you may be able to buy one of the older animals after she has kidded or when the show season is completed. Breeders like them to try and obtain their milking star*; this is awarded to goats giving over a certain amount of milk of given quality, roughly a gallon (4.5 litres), with 3.25% butterfat minimum in 24 hours. Protein is also taken into the calculation now. This award is carried on from generation to generation, and added to, thus—*I—when the daughter of a winning dam gets her own star. Some lines go back unbroken for eight or nine generations. It is therefore worth waiting for stock whose ability to produce milk is known.

The best age at which to buy is often difficult to decide, but as goats are herd animals and so dislike being alone, always try to have two. There is no need to have two milkers or two kids; they can be a mixture of any age which suits you.

IF YOU BUY A MILKER

Should you decide on a milker to give you the quickest return on your outlay, learn to milk before acquiring her. If not, you may have a 'dry milker' on your hands. This is an animal which although bought in milk, has not been milked correctly and has 'dried herself off'. Lack of stripping out makes the animal think that all the milk is not required, so she gives progressively less at each milking. Such a goat who has not yet been served will be at least five months, plus the time until service, before she is of use again.

Buying a goat who is in kid, and going down in milk prior to kidding, would do well as you would not have so much milk to remove, and your hands would not ache so much as if you had begun on the full tight udder of a newly kidded goat.

Animals also do better when they have kidded with you, rather than being moved soon after coming into milk. Thus you have the whole lactation and the kids too.

A goatling who has been served is a most useful buy. The whole of her productive life is ahead of her; there is less stress on both of you as you are used to each other, and she is also accustomed to your methods and feeding.

Lastly, kids. It's a long time before you get any return on your outlay, and you've got to put in a lot of hard work. Kids need milk four times a day to begin with (some people say milk for nine months to make a really good animal) on a definite routine. So when making up your mind what to buy, consider the time involved. Can you fit kid-feeding into your day?

These, then, are the choices: milker, or goatling mated or unmated, goatling ready for service in the autumn, a weaned kid eight to ten months old, or a baby kid.

My own preference would be a mated goatling, to produce her kids and milk in the spring, and a weaned kid, which would thus be a goatling ready for service in the following autumn. Therefore, if served alternately one each year, there would be a continuous supply of milk. But this is only my opinion, you must decide for yourself.

Prices are anyone's guess nowadays, but just consider what it costs to rear a kid for one year; you have to allow for milk (at least four pints—2.25 litres—daily to begin with and in lesser quantities for the next six months or more), concentrates, hay, straw, minerals, worm tablets, routine vaccines . . . and of course time. These add up to a lot of money. Don't be surprised when you are asked a realistic price for a well bred and reared youngster. But breeders are notoriously unbusinesslike, and would prefer their goats to go to some person who would really care for them, than get the extra £30 to £40 which might be had.

CHAPTER 2

BREEDS IN THE BRITISH ISLES

THE English goat—grey, brown and black coated with longish fringes—was thought to be extinct. However, odd animals were found dotted about the country and interested people decided to try to revive the breed. With this in mind an English Goat Breeders Association was formed in 1978. Some hostility was shown by the establishment, but this has been overcome and classes for the breed are being offered at some shows. A photograph of the English goat has been included in the new book of breeds produced by the British Goat Society.

Characteristics of the breed are being taken from paintings produced around 1900. Kids are being born with correct colours and markings, but are still inspected by officers of the English Goat Breeders Association before being included in their Register. A Herd Book has been produced.

Goats fall into two types: Swiss or Alpine and the Anglo Nubian or desert type; the former are in far greater numbers.

At the end of the last century some goats brought over to supply fresh milk to passengers on ships were of the Delta, Zaribe, Jumna Pari, and Bengalie breeds, from Egypt and India. A most distinctive type, they had long drooping ears, varying in size, but were of several colours and patterns. These goats were much sought after and bought rapidly. After many crossings, the creature has emerged with the characteristics of the original imports intact, but a much better lactation both in output and duration, as the Anglo Nubian.

HERD BOOK SECTIONS

The following list is taken from the Herd Book in which the British Goat Society registers stock as pedigree:

British Saanen Chesswood Nimue Q*2 Br Ch, owned by Miss C D Barlow, Chorley Wood, Herts, holds 3 CCs and 3 Reserve CCs.

The Champion UK Male 1988, British Saanen CH SS171/211† Ashdene Morrison is owned by Mr N J Parr, Guildford, Surrey.

Saanen. White, short-haired goat, also of direct line from Swiss goats; stocky, rather short legged, deep bodied. Excellent milker, very placid; variable butterfat, but long lactation; good winter milker.

British Saanen. Very like, but larger than, the Saanen from whom she was bred. Probably the highest milker of all breeds; butterfat very variable. Popular goat for all purposes; large specimens up to 220 lb (100 kg) adult weight.

Toggenburg. Affectionate small goat about 110 lb (50 kg) adult weight, but many specimens weigh more than that. Fawn with characteristic white marks on the face, legs and rump. Has been kept as a pure line with no out cross direct from Swiss imported stock; the last to be brought into the country about ten years ago. Has an undeserved reputation for low butterfat. Toggs give good levels of butterfat and protein, are long lactation milkers, good grazers, and reasonable winter milkers.

British Toggenburg. Bred from the Togg, but larger and darker in colour. Being larger, she usually carries a heavier milk load. Placid disposition and good long lactation. Grazes well, but will stall-feed. Adequate butterfat, very useful household milker. Great consumer of bulk feed.

British Alpine. In breeding like the British Togg, but has a black coat; face and legs have the Swiss markings in clear white. This breed looks very smart; it is a large rangy animal, very active, likes extensive grazing or long walks. Is a good heavy milker; udders not always as good as they could be. Needs a great deal of roughage to do her best. Somewhat difficult to restrain; seems to have more of the mountain than valley in her make-up.

Anglo Nubian. Quite the most distinctive of our breeds in appearance; long hanging ears, Roman nose and almond eyes. Equipped in fact to cope with the dry hard desert herbage, with ears to flap away the flies. Milks well and has the best butterfat of any breed, 5–6%. Is better served annually, or has a tendency to cystic ovaries. Colours vary from solid white, black or mahogany, through spots and marbling of mixed shades. Large but timid animal; a young kid of another kind can bully adult females. Easily restrained; as from earliest history they have been herded by nomadic peoples. Unwanted male kids very soon carry enough flesh to be a useful

British Toggenburg CH September Status Q*1 is owned by Mr R Parkin, Goole, East Yorkshire, who has many winners amongst his stock.

R120 Tyrrell Larkspur Br Ch Q*1 bred by Mrs G Mears, Chelmsford, Essex. Note the excellent head on this Anglo Nubian. (Photograph by Anthony E Reynolds)

addition to the meat of the household.

Golden Guernsey. Very small and pretty goat bred on Guernsey; is being revived by careful breeding under the watchful eye of a Trust set up on the Island for that purpose, from a nucleus of 65 animals of all ages and both sexes. Very inbred now, with inherited faults. Some animals have been brought to the mainland, and in 1987 the BGS for the first time gave Breed Challenge Certificates for them, so now the public can see what they look like. Their number has risen wonderfully well to about 2,000. Colour from deep cream to auburn, with clear mid-gold the ideal. Adult females are about 90 lb (40 kg), with yields of about 7 lb (3 kg) upward of good quality milk. Takes little space, but will not stall-feed; requires grazing. A challenge, but not a beginner's goat. On the Rare Breeds List.

British Guernsey. The breed was recognised by the BGS in 1975, but growth in numbers is slow because of the registrations involved. The British Guernsey can be achieved by using a Golden Guernsey male on a female of any breed. Female progeny from then on should be served by Golden Guernsey males, and for a further three generations female kids only may be registered. The fourth generation should still be sired by Golden Guernsey males and both female and male kids are acceptable for the British Handbook.

British. This is really a hybrid, the result of crossing two pedigrees for the purpose of fixing some valuable characteristic. Two animals, the male of which is pedigree, can be mated and the daughters produced go into one of the grading registers IR, SR, FB or HB. A pedigree male is used with each generation to improve the progeny and the offspring will go up to the next grade. British goats can be identical to one parent (pedigree alone shows the difference) or may be quite unlike any breed in particular. When crosses involving Anglo Nubians are made, these can be definitely 'one off' and unrepeatable (fortunately). As with most hybrids, they are vigorous, strong and excellent milkers.

Boer. The British Boer Goat Society was formed in 1987 when the first Boer goats were imported from Europe and arrived in this country. The breed originated in Southern Africa. After arriving in Europe, it was thought that the increasing market for goat meat in the United Kingdom would benefit from animals specifically bred for carcase production. Many members of the Society are farmers who, needing to diversify following new regulations on milk quotas etc., thought that meat other than beef, pork or lamb might be another string to their bow. Boer goats are very solid, somewhat like a short-eared Anglo Nubian goat in appearance, they grow very rapidly and produce good, lean meat in the short time of rearing.

Angora. These animals from New Zealand, Tasmania and I believe also from Canada were brought into the UK at very great expense. Since a great deal of very high quality fibre was imported yearly at equally high cost, it seemed logical to try to supply as much of the market from home-produced material as possible. Mohair, angora and cashgora fibres are available now, and a marketing board has been set up for the purpose of grading and marketing the material. Numbers have risen as further imports have arrived. Crosses have mainly involved white animals at first and subsequent crosses have produced better hair. The character of the animal is also changing. The angora goat also provides an outlet for those farmers needing to diversify.

This breed is somewhat different in form from other goats: in shape more like lamb with shorter leg bones, it carries much less fat than lamb, less cholesterol and is more acceptable to current thoughts on diet. Angoras require much more fibre and less protein in their feed than other breeds. They are pretty creatures with curly hair which is clipped twice yearly. Plenty of information about them is available. The price of stock has gone down because cross-bred animals are on sale now. Many home spinners keep a few goats for their hobby.

Pygmy. The Pygmy Goat Club was founded in 1982 to instruct those who liked goats, but did not want vast amounts of milk or have the space required for a larger breed. These small, sturdy animals make good pets, have great character, are lively, friendly and of course take less space and feeding. Like all goats they need housing all through the year, correct feeding and company. They produce less milk than other goats, but usually more than the kid requires. Preferably keep two; one could be a neutered male. They make attractive animals to have about if you do not want to be tied to milking hours. The Pygmy Goat Club registers stock and supplies information.

The Harness Goat Society. These animals are not actually of any particular breed, but are trained for a special purpose and are often, but not necessarily, a neutered male. It was thought advisable to promote correct training and to prepare 'working goats' ready to act as pack animals, to pull a travois (sled) or to be harnessed to a cart and so on. The Society was formed in 1987. The animals are trained to be led or driven, and special lectures, slide shows and demonstrations are held. Judges qualified in driving and inspection who understand the temperament and potential of the goats are asked to consider the animals' ability in showing classes. The Harness Goat Society is affiliated to the BGS and there are also societies in the USA, New Zealand, Africa and Europe.

GRADING REGISTERS

Identification Register. Purely for the use of those wanting to show or milk-record an animal which is unregistered. Female progeny or IR or unregistered dams by registered males are eligible for the SR.

SR Supplementary Register. Female progeny of SR goats, by registered sire, go into the FB Foundation Book. Also goats from the IR or SR who win Stars in milking competitions are automatically up-graded into the FB.

Foundation Book. The female progeny of SR goats sired by registered males are entered.

British. The female progeny of FB goats by registered males go into this, as previously explained. This now upgrades to British Guernsey.

There is not as much uniformity in all breeds as could be wished, some specimens being half as large again as others. Providing type and shape are correct, I personally prefer the smaller version. If she is giving a reasonable amount of milk, I can see no good reason for feeding an extra 56 lb (25 kg) of goat.

Colour can vary greatly amongst the British Toggenburgs, from pale fawn to chocolate, and all are in order except for grey fawn, called 'silver'. But the Swiss marks on BT or BA or T goats must be white, not cream or fawn, to be correct. The British Guernsey is pale or dark gold, with no white patches.

SELECT A HEALTHY ANIMAL

Whatever your final choice, the animal should be healthy. So look for an alert expression, bright, but kindly eye, and a small neat head. She should have a slim neck running into the shoulders with no coarseness, and the shoulders should not be too heavy or loose. Looked at from head to tail, the milker should be wedge shaped, getting progressively wider from shoulders to rump, with a good spring to the ribs showing her to be a good bulk eater, with a well-filled body.

There should be good width across the pin bones with the back legs sound and solid. The top line from neck to tail should be level with very little slope at the tail; a steep slope here is 'running off' and is said to denote a poor milker. That may be so, and it does usually go with a necky udder, which looks as if it were tied in at the top like a bag. This can cause chafing as the animal walks, also bruising and even mastitis if the udder is very full.

From the side, look again for the wedge in the depth of body, which should get deeper from front to back, the udder making the final depth. The udder should be attached well forward under the body, and from the rear the vessel should be firmly fixed high up under the legs, flat across the bottom and rounded at the back. Teats should be neat, of a reasonable shape and size, and not project sideways, or be like bell ropes hanging down.

In young stock, length is looked for and depth of body too. A goatling should have a lively disposition, good coat, nice pink lips and a good width between teats, which shows that a reasonable udder may be expected. Try and see the dam of any youngster you consider buying. The dam is said to control quantity, and the sire the type, so if you can see a female relative of the kid's sire so much the better.

USE THE BEST MALE AVAILABLE

Males, we are told, are 'half the herd', and using an unsatisfactory male can certainly mar your plans for years. Two years will have passed before his daughters are themselves milkers, so the best male obtainable is the only one worth using. You do not have to own him; leave that to someone with more knowledge, as a breeder would not waste time, food or money on a stud male which was not considered excellent.

With males you cannot really go on looks. So long as he is of the breeding you need, that's the criterion. Environment, too, can make an enormous difference to the growth rate and general well-being of all animals. Twins can appear very different indeed, but genetically they are not.

The bulk of stud males are kept only from milk-recorded dams. Thus evidence of his dam's performance, and that of his sire's mother, can also be studied for both quantity, quality and length of lactation.

Now that males of all breeds are found on the AI list, there is no reason why you should not use the best in the country. He could be in Cornwall or Fife but a simple request for information as to your nearest inseminator will bring the straw from the male of your choice. You will need to supply the date when your animal will be on heat. If she is not ready at the appointed time, you can ring to postpone till a later date. Straws are sold in twos in case the first insemination does not hold. The inseminator will, of course, charge for the journey to insert the second straw.

CHAPTER 3

HOUSING

THE basic requirement is a dry, draught-free house, which should be ready before you take possession of your animals. How you arrange your building depends on its size and construction, and what you intend to use it for. By that I mean, will it contain an area where hay and food is kept as well as goats, or will it simply house the stock? Also, do you intend to keep your adult goats in single pens or are they to be kept collectively, and how many animals do you think of keeping?

A stable, say, 13 ft wide by 10 ft deep (4 m × 3 m) would give you three pens each 4 ft 3 in wide by 6 ft 6 in (1.3 m × 2 m), which is a good size for one milker or two goatlings, or three or four kids to share. Pens of this size would leave a passage in front 3 ft 3 in wide (1 m), large enough to open the doors wide when

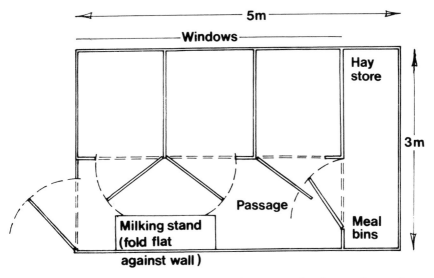

Figure 1   One way of adapting a loose box or poultry house for goats.

cleaning out, and for you to enter with food and water without having these dashed from your hands before you can get them to their correct destination.

Partitions between pens may be slatted or solid; outdoor quality 3-ply is fine as it is available in sheets 8 ft × 4 ft (2.6 m × 1.3 m) roughly. That means that with framing of 2 in × 2 in (50 mm × 50 mm) a whole pen side can be put up in one piece; it is the right height, and one rail top and bottom of the frame is sufficient to hold it firmly, so long as it is nailed closely.

MATERIALS TO USE

Concrete blocks are quick to erect and very permanent; and many of the new plastic-faced materials sold in panels, and allegedly strong enough to withstand pigs, would be most suitable as they are easy to wash down. Wooden walls are much easier to use, of course, as nails or screws can be driven in, but brick and clay-lump loose boxes are often found and make very warm and comfortable accommodation. Framework to hold any sections will probably have to be drilled with a masonry drill if brick, and holes dug into the walls if clay-lump, and the woodwork banged in and cemented there.

Floors in most of these buildings would be earth, which is not easy to keep clean and sweet. If you can lay concrete, this is first class, as you can make a channel outside your pen fronts, sloping to a hole under one wall to a drain or soakaway. The floor should slope from back to front, with an allowance of 2 in in 3 ft (50 mm in 1 m); the concrete does not need to be very thick but a good skim is necessary to keep the whole area dry. Before laying the floor, put down rubble, empty bottles, even egg trays, to give some insulation and make it warmer, as the air between the objects will warm up and help to hold off the chill.

If a hard floor is out of the question, look for the lowest part of the floor, dig out a hollow about 12 in (300 mm) deep and fill it with gravel, broken bricks or something of that kind. Replace the soil and ram it down really hard; this will act as a drain and make a surprising difference to the amount of straw used. In some places where there are chalk pits, it is possible to buy what are called chalk nobs; these are hard pieces, walnut sized or larger. Rammed down all over a floor they are excellent, really dry and warm. The only snag when cleaning out is that you have to be careful not to slip, as it is rather like a skating rink when damp;

however, it is 'limey' and discourages flies.

Doors should not be less than 2½ ft (750 mm) wide as goats get extremely bulky when late in kid, and crushing in doors has accounted for more than one kid arriving dead. Goats, being very active creatures, appear to think that speed of entry to their pens is necessary at all times.

LIGHTING ARRANGEMENTS

Lighting must be thought out too, as you will not be able to see to do the chores in the early morning or evenings for more than half the year. Can you put in an electric light? Lamps run off car batteries are safe and most useful, failing that, large torch lanterns are best. Some can be recharged off the mains, and though quite an outlay when bought, there is no risk of fire and they do last for years. All mains wiring should be well out of the reach of goats or run in steel conduit (wires are very dangerous). If no other form of light is to be found, a storm lantern is quite useful, as it can be hung up and won't get knocked over. With straw and hay about, you can't be too careful; that also goes for the odd cigarette end!

The building itself must have light inlets; in fact many have windows. However, if they can be reached by a goat standing on its hind legs, cover the windows with small mesh wire netting held in place by narrow splines of wood to keep the wire taut. This will prevent hooves knocking glass out and so stop accidents from cuts. If the window can be reached too easily, remove the glass and replace with clear corrugated plastic sheeting.

If the light area is too small, fit a corrugated translucent panel into the roof. If the roof is corrugated iron or asbestos, the panel will fit into the same hollows. If you have a stable-type door, a panel can be removed from the top half and a piece of rigid clear plastic inserted; this will give extra light. Also, this upper section can be opened, giving fresh air without chilling the occupants. Glass tiles, mixed amongst the normal tiles, will do quite a lot for dark corners in the back of the stable; many older buildings have such tiled roofs.

VENTILATION

Ventilation is a mixed blessing as it is almost impossible to achieve a warm floor and cool head area because of the warm

House for three goats on the author's previous farm, referred to below. There is a second window obstructed in the photograph by the open door.

air rising. This causes unpleasant down draughts. If you feel that more ventilation is a must, a louvred area, baffled on the inside and placed about 6 ft (1.8 m) above floor level, will do well. Hopper windows are a help too, but as they hang inwards they must not be behind the pens as they take too much space. Sliding windows behind the pens, as found in some sectional buildings, are excellent as they can be opened on both sides of the building, but see that they are not directly opposite each other or you'll get problems with draughts.

Many firms make good sheds in a variety of sizes. If putting up a house specially for your goats, and having decided how many goats you intend to keep, say, two milkers for your household, always remember that that means at least three pens. Why? Well, when one or other of your goats kid, the kid has to be put somewhere, as you won't get very much milk if you leave the kids with mum.

I have a good strong elm ship-lap shed for my three males. It was supplied by a firm that was willing to pander to my needs at no extra cost. I therefore had it made 10 in (250 mm) wider, overall 13 ft × 9 ft (4 m × 2.75 m). It has a pent roof, that is, sloping in one direction. There are two windows 2 ft (600 mm) square, one at each side of a central door in the 'high front' side; the door itself is made in two parts. There is space at the end of the passage for a bale of hay at one side and straw at the other end. This saves many steps on bad days. The pens are 6 ft 6 in × 4 ft (2 m × 1.4 m).

In high buildings, a false ceiling, which can be of wire netting held by slats and nailed on to the beams, and then covered with bituminous paper or even split mealbags, will stop cold winds blowing through tiles, and make an astonishing difference to you and your herd.

WHERE TO MILK

As it is not a good thing to milk where the animal lives, you should have some space where you can bring out the milker. You can put her in an empty pen, unlittered, or at the end of a concrete passage outside the pens, or indeed in a completely different building. However, unless you have a large number of goats, this can waste much time and space. So for a small herd, use a milk bench, which can be placed outside the penning area.

Milking benches are most useful as the goat is otherwise rather near the ground and not so easy to get at. If the animal is standing 10 in (250 mm) above the floor, for milking you can either sit on the edge of the stand, or on a stool and be more comfortable and see what you are doing. Made of wood, about 20 in (500 mm) in width, the bench can be hinged on to the wall at one end and have two strong legs, also hinged, at the other end. When not in use, the legs will fall back against the under part of the bench, which will then swing back on to the wall. Fastened securely with a hook and screwed eye, the whole thing can be let down into position in seconds and you will not be always falling over it. If you have plenty of space, a concrete block in a milking bail is suitable and easy to swill down and keep clean.

Each pen should be fitted with a hay rack, not nets. Goats have a nasty habit of getting their heads or feet into a net, and can thus injure themselves. Racks may be constructed of wood, with solid ends and rails not more than 2 in (50 mm) apart, or they may be of weldmesh, 2 in (50 mm) squares, on wooden frames. It is possible to buy racks made to slot over a partition, leaving one rack each side; these are metal framed and mesh covered, and are very good indeed. Though made for calves, they are ideal in size,

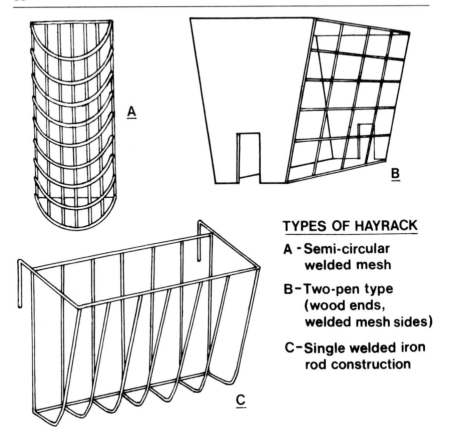

Figure 2   Three types of hayrack.

being roughly 20 in (500 mm) wide and of the same depth; they measure 14 in (350 mm) from back to front and slope, of course, inwards at the base, about 8 in (200 mm) there. It is not difficult to make the same for yourself in wood. Sometimes it is fairly easy to get a half-round, flat-backed rack, again of weldmesh or similar material; the bottom is half-moon-shaped, floored with a piece of plywood. Front to back it is 12 in (300 mm) about 16 in (400 mm) wide and 28 in (750 mm) deep. These racks can be fixed with staples, or indeed tied on to a rail or two hooks, and are especially good for kids of several ages together as they can eat at whatever height they can reach. Taller or shorter kids can reach below or above and as the racks are rounded in front there are no corners for small ones to bang against.

BUCKET HOLDERS

A shelf with a hole cut in it will hold a bucket for food or water. A ring of heavy wire, hooked into two eyelets so that it falls flat against the door or wall when not in use, will serve the same purpose. If pulled straight out and the pail inserted, it stays put. This type of holder is often screwed on to the outside of the door, through which a hole is cut; the animal then puts its head out to eat. The food pail is replaced by the water pail as required. The hole must be at least 10 in (250 mm) wide × 14 in (350 mm) deep, preferably in a keyhole shape. The head is put in at the widest part and lowered into a narrower section about half the width of the top, that is 5 in–6 in (120 mm–150 mm); the ring should be about 10 in (250 mm) above the floor.

By using a short length of chain or twine fixed to the pen wall, water pails can be hung so that they are not tipped over. The old fashioned stretch type of curtain rail placed across an angle across a corner, will stop food pails from being taken right to the back out of reach. Always place hayracks either on the front rail or on the dividing wall between two pens, right at the front where you can fill them without entering; opening and closing and bolting each door is a sheer waste of time.

FITTING BOLTS

Bolts are important too, as goats being intelligent or crafty— however you regard them—are able to open an assortment of latches. The self-closing latch for garden gates and so on will only do if you drill a hole through the rounded flap and thread through it a nut and screw on a bolt, otherwise the animals will depress it and walk out. Latches with down-pressing handles are useless, for the same reason. A good strong bolt, of the type which has a hook on the end of the handle and turns under when closed, is fine, but don't place any bolt at the top of the door (goats have long necks)—put it at least 14 in (350 mm) down.

People with larger herds in mind will do well to get and study the catalogues of firms making deep-litter poultry houses, which come in sections and are simple to erect. Ends are usually fitted with sliding doors, which are very draughty, so ask for swing doors to be supplied. As they are usually 4 ft 3 in (1.3 m) wide, have two doors fitted with hinges top and bottom, otherwise the weight will make them sag. Windows go along the entire length. As these houses are of standard width and have even-sized sections, many layouts are possible.

I have such a house myself, which includes ten milkers' pens, one double sized for kids or goatlings, and another for goatlings or kids, not quite so large. With a central passage 5 in (1.5 m) wide, cleaning out is simplified as a barrow can be got in easily. A gate across the passage two-thirds down keeps the animals their side, and the milking bench is let down across the gate when required. A tap for watering is also in the passage.

At one end of the house is a section in which twenty bales of hay and ten of straw are kept; there are also three large metal feed-bins for different cereals. Walled off completely, but under the same roof, is my dairy which has a sink with cold and hot

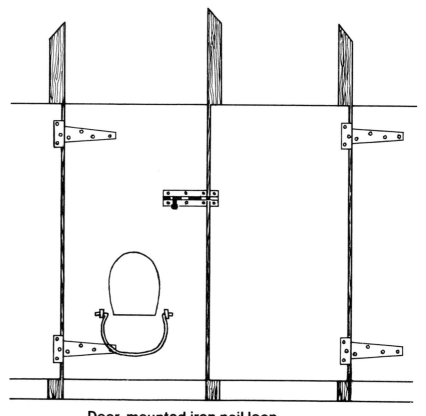

Door-mounted iron pail loop

Figure 3   Feeding arrangement for single pens.

Echin Annabeth *4 BrCh is a very nice type of Saanen. Bred in Hexham, Northumberland by Mrs Marion McKechnie, she now resides with Mr Tony Smith at Farnham in Surrey.

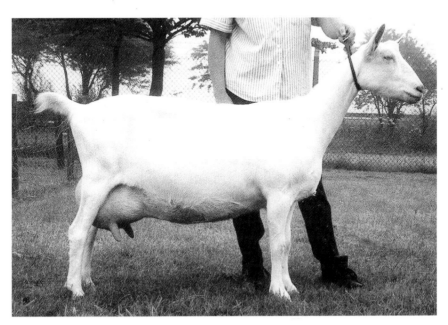

A good example of a British Saanen, CH AR271 Toddbrook Marrakesh Q*3 BrCh manages to encompass high yields with excellent conformation. Owned and bred by John and Hilary Matthews of Chelmsford, Essex.

The Toggenburg one would hope to breed. R148 Meadowlark
Rosamunde *2 BrCh, owned and bred by Mrs Joan Archer of Dorking,
Surrey.

RM202 Tetherdown Charisma Q*4 BrCh, a fifth generation British
Toggenburg female breed champion, has already produced two champion
daughters. Bred by Mrs Ruth Tyler of Takeley, Essex.

An excellent animal in all respects. Tyrrell Swish Q*2 BrCh is owned and bred by Mrs Gaye Mears of Chelmsford, Essex.

A really lovely British Alpine, CH RM185 Nebo Natasha Q*2 BrCh, showing all the good points of the breed. Owned and bred by Mrs Sally Williams of Stowmarket, Suffolk.

A winner of many prizes, R116 Kerney Naomi is a very good example of a Golden Guernsey. Owned and bred by Mrs Linda Mudle-Small at Golberdon in Cornwall.

Pygmies have become increasingly popular in recent years. These two attractive Beckvale kids were bred by Mrs Joy Wolton of Diss, Norfolk.

water for cleaning utensils, benching and tables, metal shelves for cheese presses and so on. It also holds a large 30 cu ft (0.85 m^3) freezer, in which the milk is frozen in 1 lb (4 kg) blocks moulded in polythene boxes. It is a very self-contained unit.

The same type of building would do splendidly for anyone who wanted to have a collectively kept herd; this means a largish area for living and some very definite method of seeing that every member gets her rations. Bullying takes place in this type of arrangement unless the group are all one family, and even then there is a 'boss' animal which has first go at hay racks and green food before allowing the more timid animals a look in.

FEEDING ARRANGEMENT

One type of feeding arrangement can be constructed by fitting a wooden paling across the building, with pales fastened to a single rail at the bottom and between two rails at the top, one at

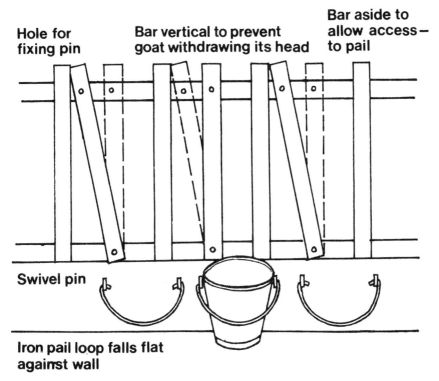

Figure 4   One type of feeding arrangement.

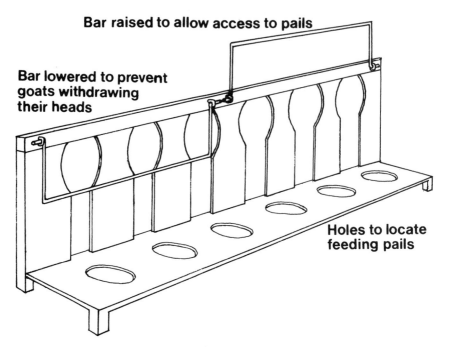

Bar raised to allow access to pails

Bar lowered to prevent goats withdrawing their heads

Holes to locate feeding pails

Figure 5   Alternative form of restraining collectively-housed stock during feeding.

each side of the end uprights. One pale, pivoted at the bottom by means of a nut and bolt, is drilled at the top to match holes made in the two top rails. By placing a pin through all three pieces of wood, you are able to secure the pale, which can be swung aside to allow the goat to put out her head and fastened upright again when she has done so. A ring to hold a food or water bucket should be fitted in front of each opening (see figure 4).

Alternatively, you can secure the animals by lowering one long bar across their necks along the whole length of the building, fastening the bar at one end.

Another method, which involves individual handling of each animal—and is only practical where small numbers are to be kept—is to fasten each one separately to a stanchion along the wall with a short chain and collar whilst they are feeding, then release them. Ample hay racks must be installed so that frightened or young animals are not deprived of their needs. Hornless and horned animals must not be kept together in these circumstances as damage will be done to some member of the herd.

The exit to their yard or pasture should normally be at the opposite end to the feeding area; this leaves a clear side for hay racks.

Water pails of course replace some of the food pails when the latter have been removed. About twenty minutes is usually sufficient time in which to eat the concentrate ration, though some animals may take much less. None should be released whilst any member is still eating, or those released will immediately worry the remaining animal who is fastened—so one out, all out!

FITTING HAY RACKS

Hay racks can be arranged above the front feeding area, which would save going into the pen, and they should be fixed high enough for the goats to stand and eat from the lower edge. If the racks are too low, when bringing out their heads, the goats lift them and so bang them on the base of the rack. Make sure they are at least as high as the head of your tallest animal; smaller ones will in any case like to stand with their feet on the front rail, as is their habit.

A floor area of 9 sq ft (0.84 m²) per animal is needed, as goats in winter spend much time indoors in a collective house, which is more or less a covered yard. You will need a gate into the milking area and this can be the same width as a pen door, 30 in (750 mm) approximately. The outer door should be wider for mucking out. Large double doors are more draughty, but do allow a tractor in for an annual clearance.

Allow at least 3¼ ft (1 m) of head room more than normal— 6½ ft (2 m)—as there will be a build-up of at least that amount of bedding in a year. Goats are very wasteful of hay, pulling out more than they can eat. Once down on the floor it is wasted, so put lids on your hay racks—mesh or solid, it does not matter. I have used baker's trays, refrigerator shelves, oven shelves and all manner of things as I hate waste, especially as hay is now so costly.

SITING YOUR HOUSE

Try and place your house facing south or south-west. Also a good path will facilitate the delivery of meal, hay, straw, and the movement of animals. It will also be cleaner; less mud is carted into the building to soil the straw in a small house, or the passages

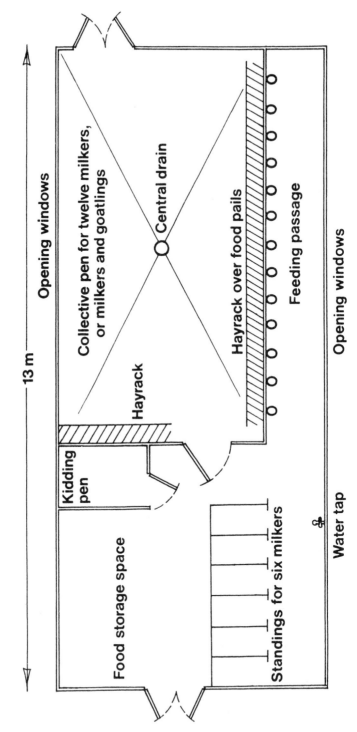

Figure 6   Plan of collective goat-house.

in a larger one. Instal mains water, if possible, or otherwise have a water tank outside to save endless walks with slopping pails. A few yards of alkathene piping with a tap screw on one end fixed to an outside water-point, if you have one, would be a great help.

A concrete apron outside the doors will prevent mud being carted around and also animals or humans slipping.

Large railway vans, made of heavy wood, are sometimes advertised and being roughly 20 ft × 10 ft (6 m × 3 m) could be made into useful housing, with adequate window space fitted. These vans are very strong and have large sliding doors at both sides. The main difficulty is their weight and getting them delivered to the home.

At sale yards you can often find heaps of wooden sections from the very large crates for machinery. These offer a number of ideas for the handy person to make up into buildings.

Finally, if you live in an urban area, before buying or erecting anything check with the local authorities that there are no bye-laws stating that you may not keep livestock or erect a building over 5 ft (1.5 m) long, or something of that sort. If you are considering a goat house larger than 150 sq ft (14 m²), you must inquire of the Planning Department of the Ministry of the Environment as to where you can place the building. You may well have to submit plans in triplicate and await the planning authority committee meeting, when you may be told to alter shape, material or site. So don't rush in, inquire first.

ELECTRIC FENCING

Electric fencing marketed under the name of Flexinet is a great advance upon its forebears. It is fitted with seven lines of electrified wire going lengthwise and polystyrene twine vertically, and it is made in squares. Even tiny kids do not argue with it; if they touch their noses, they yelp and turn around, and then immediately sting their tails. After that, they are very keen to have nothing to do with it; in fact you can leave it off for hours and they will never know because they do not try to touch it. It is very light and has fine built-in plastic rods for keeping it upright. It really is quite the best kind I have ever used, and I have tried most kinds during the 40 odd years I have been dealing with goats. It is really saying something that now I use only Flexinet for inner and outer fencing and for dividing paddocks. My fence is run from the mains electric supply, but it can be run just as well from batteries.

CHAPTER 4

FEEDING

A GOAT'S feed requirements comprise a mixture of bulk foods needed in great quantity for her metabolic use. She has an enormous potential for using roughage as a means of making milk; her capacity relative to her size is far greater than that of cows. Plants such as brambles, hawthorn, ivy, thistles and nettles, which would not be eaten by cows or sheep, are greatly liked and greedily eaten by goats. Even small kids will tackle them, though their mouths appear so soft. Use should be made of this ability to use roughage; the goat which is stall-fed or yarded will astonish and slightly horrify her owner by constant demands for more.

Let's take the more usual food first, as it is the most easily fed and quite definitely a necessity: good hay. If you could have one food only, this would be the one. Clover hay, or clover with cocksfoot, is best without doubt. It is not easy to get a good sample, which is so important. It will be unacceptable if damp, going fusty and dusty, and will simply be wasted. In spite of anything you may have heard to the contrary, goats are unbelievably fussy about their food. Lucerne hay is very well liked, but again is sadly in short supply, which seems strange when it is considered that the crop will stand for several years without being renewed, and more than one cut can be taken in each year. It does have to be kept clear of weeds to begin with, but once established is not difficult to maintain.

LEY MIXTURES

There are many ley mixtures which, providing they are indeed mixed, are excellent. However, a lot of monoculture is practised today and consequently the mixtures do not give sufficient variety for goats, to whom variety means vitamins as well as the spice of life. Ryegrass and white clover, loved by pony pasturers,

Are the Chesswoods dead or only sleeping? *From front left to right: Daisy, Kittijane, Nimue Q*2 Br Ch, RM209 Jade Q*1, R243 Sally Q* (behind tree), SS199/320 † Pamart Ferrari, Merillion (1 CC and 2 BCCs).*

are useless for goats who will only eat ryegrass on sufferance and entirely ignore white clover. Good meadow hay is really preferable to hay from leys as it contains many herbs. But if putting down a mixture to cut later, it is well worth getting one which contains coarse grasses and herbs; these are known as herbal strip mixtures. Hay, as I have said, is a must. Reckon on 10 cwt (500 kg), for each goat or goatling, and a lesser proportion for a kid, up to 5 cwt (250 kg) for the year; it may not eat this amount, but in all probability it will.

You can at times get good pea-straw baled; that left behind the viners for the processing trade is not much use as it is soiled and short. The peas intended for drying are much better; the haulm is nice and clean and long, and it is often quicker for the farmer to bale and move off the field so that the next cultivations can be started, than to burn it. Buying in bulk, directly from the field, is cheaper too.

Pea straw can be fed at least once daily and will help make the more expensive hay last; it is very well liked, being in texture like the clover, coarser and crunchy.

Bean straw is also to be had at times, and like pea straw is of use but not so well liked. I use it for males at times, as they are not so fussy as their ladies. If it is pulled out and used as a bed, it does not break my heart to quite the extent that good hay being so used does!

Oat straw has fair feeding value, and goats will eat some, but if filled up with it, they will have less space for more nutritious green weeds which are available at the same period. So the rule is—*not too much*. Being very contrary creatures at times, if clean straw is laid for their beds, goats will invariably eat quite a lot of it; put the same straw in their racks and you will get withering 'do you think I will eat that' looks.

GREEN BULK

Green bulk is basically provided by various kinds of kale. Marrow stem kale is used up to Christmas, after which it is usually split and spoilt by frost. This is followed by either thousand-head kale or *Maris Kestrel*, another branching variety but not so tall; this is most useful in windy places.

Ox-heart cabbages can either be cut in halves and placed in a pail to be scooped out, chopped, or hung on a cord; so it provides entertainment as well as food. Kohlrabi, Jerusalem artichokes, swedes, turnips and, in fact, all varieties of brassicas are widely used as bulk feed. Chicory, the field variety, is said to be the only green food which supplies all the minerals a goat needs in one vegetable. Added to that, it is very useful in time of drought as its roots go down to great depth. Always sow some in your grazing area.

Tree branches, though cumbersome to cart, are very useful; they add variety and the bark is excellent for digestion. Goats are really browsers, not grazers, and need hard herbage of some kind. A branch tied up in a pen will also help the young animals to shed their baby teeth. (They shed two teeth each year and grow two permanent ones in their place. The full mouth has eight permanent incisor teeth at four years.)

Branches will also prevent bored animals from gnawing the woodwork. Trees and hedges are their first choice, and that is why goats have gained such a bad reputation in some parts of the world. Kept in far too greater numbers for the area, they soon make it into a desert by eating the bark and small branches off the only standing trees. These naturally enough die as a result.

Then there are weeds. You can take a hook and cut enough to fill a large rack in the goat yard in a few minutes. Many otherwise useless plants are put to good purpose. Sheep's parsley, campions, yarrow, dandelion, willow herb and hawksbit are amongst those found around the average ditch or hedge in the country. In towns, you may well have to do a bit of scrounging from the nearest—or most obliging—greengrocer who has vegetable trimmings which the housewife does not appreciate paying for. Cabbage and cauliflower, celery and turnip tops, damaged apples, even bananas, are liked by a lot of goats.

Beet pulp, either dry or soaked, is a very good addition to the bulk feeds which are so vital. Though many people feed it dry, I think it is slightly risky as it swells so much when damp. Imagine the fluid taken out of the system by soaking all that dry bulk. For an in-kid animal the bulk intake of soaked pulp is known, but the space taken, when the swelling is internal, might be too much for her. Beet pulp must never be soaked in any galvanised receptable, as the acid produced will remove the coating and poison the animals as well.

FEEDING SILAGE

Silage is little used for goats; some animals will eat it, but it is difficult to make in small enough quantities. On the market at one time were some very heavy plastic bags called Mowbags; these were for the making of silage, which could be left *in situ* and used where made. This was fine. I tried some myself, but the snag was that the bags, once the silage was made, were too heavy to move, and the goats were in their house during the time I wanted to feed the silage. Silage is not the nicest smelling material to handle, particularly if fed indoors. Badly made silage can cause listeria which must be avoided at all costs.

Next time I used smaller bags: fertiliser empties, filled with well-wilted green stuff, and pressed. This was done by placing boards over the sacks and taking the garden roller across them many times. I drew out all remaining air with the vacuum cleaner hose, wire-tied them, and wheeled them on a barrow to an empty calf pen beside the yard. There they could be emptied into a bin for the goats to help themselves; it worked well. If the bags are left out in the paddocks, the goats love to jump on them and their sharp hooves cut the plastic. So even though they are weather-proof, cover them with something solid.

For the silage I used the cut herbal mixture, which includes chicory, sheep's parsley, agrimony, burnet and rib-grass; however some of the plants have hard stems, so the greenstuff must be used young and not allowed to get woody. All must be chopped into short lengths.

Lucerne cobs or grass nuts are useful. These cobs are supplied in several percentages of protein: 12–15%, 15–18% and 18–25%. Added to a winter diet they can be used to bolster the protein, but they must be worked in and not just added to the whole, or too much protein can result.

Grass nuts help butterfats, we are told, and they are liked by most animals. Usually they can only be bought in large lots, but many goat clubs buy several tons and then let their members have a share; but you must state your needs prior to the order. If you are in the neighbourhood of a mill which handles these nuts, no doubt you could buy in smaller amounts.

I believe it is still possible to buy distillers' and brewers' grains. As these are wet they will ferment if kept long so use within a short time. They are a useful feeding material.

Haylage or horsage, a silage made in large polythene bags from partly wilted grass, is quite well liked by goats, but once opened it will go mouldy quickly, so only buy if it can be shared between a fairly large number of goats, other stock or with another goat-keeper.

CEREAL FOODS

Cereals are often referred to as the 'concentrates' as opposed to the 'bulk' previously described. Bulk usually furnishes the maintenance, and cereals the production. Very high yielders need an extra pound (450 g) daily for body building; there is simply not sufficient nutritious bulk for their metabolic requirements.

Bran, which contains the outer husk of wheat, is an open and slightly laxative food. Its make-up is such that it can be added in any quantity to balanced mixtures without upsetting their balance. It is right for milk, but one could not expect any animal to like just one type of food alone. It is used as the basis of many mixtures, contains phosphorus and is good for assisting bone formation; easy to digest, it is suitable for all ages of goats and is particularly useful immediately before and after kidding.

Goats prefer mixtures as coarse as is practicable. They will not, like cows, eat endless quantities of dairy nuts; some may eat a few, but the majority nose them out of the pail and they are wasted. It is therefore less expensive and less annoying to begin with something they like.

The idea that the meal must add up to four parts of carbohydrate to one part of protein is straightforward enough, but how to compile a suitable feed is another matter. If you are like me and go to the miller and ask for a mixture to include whatever you want, and make it a 4 × 1 please, he, bless his heart, does it for you. Easier still, buy one of the excellent 'coarse dairy mixtures', 'coarse calf rearers' or 'coarse lamb and ewe' foods which all the well-known manufacturers make, then dilute it with bran, as the goat likes a lot in her pail. The heavier mixture looks very little on its own. One 56 lb (25 kg) bag of coarse dairy mixture and 28 lb (12.5 kg) of bran mixed would give you feed which would do very well for your milker. Anyone using calf rearer, which is higher in protein—56 lb (25 kg) mixture, 28 lb (12.5 kg) bran and 28 lb (12.5 kg) crushed oats—will find it will do the same. The oats bring down the protein content to one much closer to your needs; this mixture should be fed at 3½ lb (1.5 kg) to each gallon (4.5 litres) of milk produced—less for lesser volumes of milk given. Crushed oats are universally liked by all ages from small kids to adult males, and all are safe on them.

Rolled barley should never be more than a quarter of the ration, as too much can cause a type of poisoning; it is very heating. Young stock make good use of it, especially in cold weather; it puts body on to them and improves condition. It can be mixed into the milkers' food, and is well liked. Coarse middling can be used too, but it tends to be a bit pasty in the rumen and is not well utilised by goats; still, in small amounts, it does help to ring the changes in the diet. Maize is usually flaked; all animals seem to like it. It is much used in mixtures, but it is very expensive for the amount of good it does in food value, so use sparingly. It is very warming in winter, but reduce the quantity in summer feed. Sometimes it is sold as kibbled maize, which older goats like but kids cannot digest.

For the very heavy milker (11–15 lb (5–7 kg) daily), coarse dairy mixture can be fed for three parts of the ration with just under one-third of bran, and about one-eighth of dairy nuts added if she will eat them. Animals giving these yields, and up to 18–22 lb (8–9 kg) daily, simply cannot cope if it is all bulky food. If you can do so, it is better to feed their concentrates in three feeds rather than the usual two.

Dried beet pulp can be added as one part of the ration for low yielders and goatlings, but not to in-kid or very heavy milkers. Males must never be fed beet pulp as it can cause them to go sterile; mangolds have the same effect, so no mangolds for them either. Apart from that they can have the same food as the females. Males do not need a production ration but must be well fed; their meal ration during the rutting season should be greater than in spring and summer. It is hoped that they can go out and fill up on grazing, with a 'goatling-sized' meal ration, that is, ½ lb (225 g) am and pm, hay ad lib and any green food such as hedge trimmings. Kales, cabbages and chicory are also good for males. Soaked beet pulp is, of course, perfectly satisfactory for milkers or goatlings in kid; just don't feed it dry.

SUITABLE RATIONS

(all parts by weight)

The following mixtures which can be mixed at home may be of help; they can also be made up by the miller. They are suitable for animals giving 3.5 to 4% bf.

Ration 1

2 parts crushed oats
1 part flaked maize
2 parts kibbled beans
1 part linseed cake

Ration 2

2 parts crushed oats
1 part flaked maize
1 part wheat bran
1½ parts decorticated ground nut
  cake

Ration 3

2 parts bran
5 parts crushed oats
1 part linseed cake
1 part kibbled beans

Ration 4

2 parts crushed oats
2 parts coarse middlings
2 parts kibbled beans
1 part flaked maize
⅛th part molassine meal

Ration 5

4 parts bran
4 parts coarse middlings
8 parts flaked maize
4 parts decorticated groundnut cake

Ration 6 (for low yielders and young stock)

2 parts flaked maize
2 parts beet pulp
1 part crushed oats
2 parts decorticated groundnut cake

The protein part of the feed does not have to be decorticated groundnut cake or linseed cake; beans have been mentioned, but split peas, soya meal and grass and lucerne meal and nuts are all good—you need slightly more, except for soya which is high in protein.

ADD MINERALS

None of the home-mixed rations given on this page contain the minerals needed by the goats, and which are incorporated in proprietary brands, so you must either add them or have them available for the animals to help themselves. Many firms make special mineral additives for goats. These should be added in the suggested quantities. You can, however, get licks (small ones which contain a very good selection of trace elements and salts); one of these can be hung in each pen. There are larger blocks, about 28 lb (12.5 kg) each, which are called Rockies. Some licks are heavier in one or other mineral. Goats need cobalt in quantity to enable them to take up calcium. Large cobalt blocks are usually termed 'sheep'; the red blocks ('cow') are magnesium and are necessary especially in the spring when grass is lacking in this mineral, causing grass tetany or hypo-magnesaemia. Usually I have several cobalt, magnesium, iodised and glucose blocks hanging where all animals have easy access to them at some part of the day; they are impervious to weather, so putting them in the grazing area is quite a good idea. I have three types—white (usually referred to as 'yellow' by the manufacturers and marketed as Stoclics')', red and iodised—hanging in the milking area, where animals coming in or out can stop and help themselves. Prior to kidding, the cobalt blocks are gnawed into holes very quickly, but for a small herd they would last a long time and prove of great benefit.

The amount of feed to give exercises the minds of many people when they begin to keep goats and no one appears to be very definite about it, so I will give some suggestions for the normal goat.

Very few people are fortunate enough to have woodland, unlimited moorland grazing, or heath and hillside, suggested in one excellent work on goats. We will assume the average goat-keeper to have about half an acre (0.25 ha) which, divided into two paddocks and used alternately, would give a good deal of food for say two milkers, one goatling and one or two kids from April to September. Count the grazing from July onwards as filler and entertainment only; the early protein will have gone with the first flush of grass, and the later growth has little food value.

The goats would all need hay throughout the year, minimum

4½ lb (2 kg) daily for each milker and less for the youngsters; if
the hay is good and there is mixed grazing, this would count as
the maintenance ration. One milker, newly kidded, giving say 9 lb
(4 kg) daily, would need 3¼ lb (1.5 kg) of concentrates fed in two
halves. The second milker, running through (that is, milking in
her second year) giving 7 lb (3 kg) should have 2¾ lb (1.125 kg),
the goatling 1 lb (450 g), and half of that amount i.e. ½ lb (225 g)
for the kid; half should be given in the morning and half in the
evening. The kid's ration would need to be increased in her second six months (when the milk feed is being decreased) up to
1 lb (450 g) daily. The goatling, once she is in-kid, would also
qualify for an increase gradually to reach the milker's quantity at
two weeks prior to kidding. The week before and the week after
kidding, at least half her ration should be bran; this is even more
necessary afterwards than before. Too much protein at that time
may cause milk fever, so be careful to have some bran handy. The
sudden upsurge of milk can deplete the system of calcium, which
causes the trouble. That is why goatkeepers do not steam-up goats
at the last minute.

Male kids need more food than their sisters, as they have to
begin their working lives at six months. They should have the
same amount of food as goatlings, 1 lb (450 g) daily, and that
should be increased to about twice the amount gradually as the
goats start coming into season. At that time some young males go
off their food for a while, and thus a good start will prevent them
losing condition and energy just when it is most needed.

WASTE MATERIALS

Where various waste materials such as bread or biscuit are to
be had, they can be added to your list of extras. They provide a
good filler, which is eaten ravenously, but don't allow excess—it
is said they can make an animal liverish. At one time shortly after
the Second World War, I knew of a goat who had lived her life in
the yard of a London baker, who also made fruit pies. This goat's
diet was continuously bread, apple peelings and cooked waste
pastry. She thrived and milked well. Having been bombed out,
she was brought (by taxi) to the country, but was never persuaded
to eat that nasty dirty green stuff growing in the fields. Bread or
nothing, she lived for years—a remarkable animal.

Some canning factories do have pea pods, broad bean pods

and carrot trimmings; if you live close enough, you may be able to collect some of these waste materials.

Stock-feed carrots, beet, cabbage and mangolds are sometimes to be bought. Mangolds are really a farm crop, and one always hopes to find a farmer who will sell some in less than 10-ton lots! They should never be fed to stock before Christmas as they are not ripe until that time; also, they must never be fed to males as they can cause sterility and stones in the urinary tract. Too much sugar, beet pulp and sugar beet have the same effect.

Orchards are a good source of food; many of the 'pick your own' fruit farms will let you buy the dropped fruit which lies in masses beneath the trees. Ask how much for the windfalls, and you will be agreeably surprised at how little is asked for them. They will keep a long time, and if cut up, all the bruised parts can be left out. Feed about 1 lb (450 g) daily to each milker; more depresses milk yield and causes scouring in young stock.

Beet tops from the fields can be fed very well wilted. Comfrey can be cut and fed; artichoke tops and tubers later.

Fruit bush prunings, rose prunings and dead heads are all grist to the mill and are relished. Hedge trimmings and felling of trees does give a lot of food material. Be careful to remove all twigs from blackthorn, bramble, hawthorn and roses as soon as the goats have cleared the leaves, as the spikes can pierce the udder which is very soft and tender. Thorns can do untold harm, carrying in bacteria; consequently mastitis could easily be the result. These branches are best fed in the kind of outdoor rack used for ponies or bullocks at pasture. The racks, however, must be really strong, as young kids will almost certainly use them for a gymnastic display and jump in and out, so no nails or spikes should be left sticking out anywhere.

ZERO-GRAZING

With zero grazing, that is, no grazing whatever, all food is carried to the goats. They will need the same hay and concentrates, but all the additional bulk will have to be supplied in various forms.

It is essential that you have a small exercise yard. This can be concreted and kept swept and tidy—so necessary in an urban area. Manure can be composted and the more strawy part burnt. A rack made of rustic wood can be either stood in the yard or hung outside and very bulky food, hedge cuttings, prunings and branches placed in it. This will ensure that no prickly matter gets into the animals' bedding. It is good also for hooves to be on a hard surface for some time, as it helps to keep the foot in good shape and cuts out the need for trimming quite so often. Young kids love to jump on and off something, and a box placed closed side up, or a large log, will give hours of fun.

Ivy, without its berries, is a good tonic, and a goat who for any reason is a bit off her feed will often respond rapidly to a good bunch of ivy hung in her pen. This is the best way of feeding most kinds of green material such as kale and cabbage; just tie a string round the stalk, and let the animal bite off what it wants. Be sure to leave no loop large enough for the head or foot to be entangled.

Since moving house I am now keeping goats in stalls and yards. Unfortunately I have found that the graze-reared goats I brought with me would not settle to this method, so I now rear young stock which have never grazed.

My goathouse contains four milker pens; one for goatlings and a larger pen which can be divided for two age groups of kids. The partition can be removed to provide a further larger pen. There is a passage with a drop-down milking stand and a stable-type door into the yard. This is paved with concrete slabs, surrounded by 4 ft high chain link on 5 ft concrete uprights, with three wires to hold it in place. The pen divisions and doors are heavy weld mesh on angle-iron frames. There are bucket holders on the gates which drop flat when unused. The yard is twice the size of the house and contains a large wooden covered rack with a manger. Pea straw, all types of grass, twigs etc., are fed in this rack.

The goats are milking quite well, two recorded last year both gave 1,300 kg, which is quite a lot of milk from an animal that size. I only have Toggenburgs and Golden and English Guernseys. As they have all given much the same amount of milk, the feeding must have the same effect on all three small breeds.

However, it is hard work to keep any number of goats this way. All the bedding has to be carried in and all the manure disposed of over 4 ft fencing and then stacked neatly so as not to be an eyesore.

CHAPTER 5

BREEDING

BREEDING can be a science or a matter of chance. When you begin goatkeeping, you have little idea of what potential your goat possesses, and until you have had some time in which to get used to handling, feeding and coping generally, you will not get the best possible result from whatever animal you have. So do not be too ambitious at first. Aim for a sound, healthy animal who will give you a reasonable amount of milk. If she is so heavy a milker that you don't know how to look after her to keep up her yield, you will lose the milk, or the animal will lose condition from 'milking off her back' and that cannot go on indefinitely. So try and obtain a good medium-priced animal, milk her for a season, then find a male whose dam gives more milk than your own animal and is good in those characteristics where your own is poor.

A steep slope to tail or a goat who walks narrow, or has 'cow hocks', can produce improved offspring if you use a male whose dam has good hocks, a nice level top line with a slight slope at the tail and so on. Thus your aim would be for the kids to be better than their mother.

POLLED STOCK

The hornless factor causes problems at times. It is quite natural that you should like the idea of having hornless kids, but the use of two polled parents is taking risks. Polled animals frequently also carry the genes for intersex or infertility. The bisexual animal, which can have almost complete sets of organs for one or indeed both sexes, looks like one and frequently acts like the other. It cannot breed, but as it so closely resembles one sex it is often kept until breeding age before it is found wanting. This wastes time and food. Also, the owners get attached to their animals and hate having them put down, however useless they are. So be wise and use one horned or disbudded parent.

Occasionally a kid, apparently female, will be born with a small swelling at the base of the vulva; this is definitely an intersex and should on no account be kept.

Not all polled animals carry this gene, and if you find one which has a horned grandparent it should be quite suitable for breeding. Also, not all kids born to hornless parents are hornless themselves if there are 'horns' in the ancestral background. Those males would be right for either disbudded or polled females. You should if possible find out the milk quantity given by the male's dam, also her butterfats. Inspect her, or any female relatives of the male. This will give you a good idea of what your doe kids will be like as the sire rules type and the dam production. It does not, of course, always work this way, but it is a rule to keep in mind.

STUDY MILK AND FAT RECORDS

Recorded herds will have all the milk and butterfat information on their record book, but in other cases you will have to look at show wins and so on. Most large shows have milking competitions; these will give you the yields and butterfats in a 24-hour period. Any goat giving a specified quantity of milk and fats gets a Star, which appears beside her name in the future. As the conditions are different from those at home, it is a good test of the animal's worth, but a nervous beast is at a disadvantage. Protein is included in milk recording and competition milking, so it must be looked at where possible.

Line breeding is practised by many breeders, when one particularly good animal is used more or less as a pivot, being found on both sides of the pedigree. Grandsire of your female is father of her future intended, which will produce progeny carrying the line from both parents and so on; that doe in turn could be mated back to her sire's brother or her grandsire.

In-breeding is very complex and not to be considered by the unknowing. It means possibly dam to son, sire to daughter, brother/sister matings, with a further cross back again in the grandchildren. But there must be an outcross in between, otherwise all the weaknesses as well as the good points will be fixed.

Obtain a stud goat list from the British Goat Society; issued annually it gives the names of all males in the country and is also split into breeds in areas. Failing that, a list from the local goat club will give you details of animals closer to you. Study it and, as I have said, go and see any that seem suitable.

CH AR 213 Ashdene Ming Q*3 Br Ch, owned and bred by Mr N J Parr, Guildford, Surrey. It would be difficult to find a better British Saanen.

COMING ON HEAT

When September arrives, watch the goatling or milker you wish to have served, and when on entering the goathouse you hear a tapping, which is a tail banging against the partition, or see a constantly wagging tail, look at the vulva which will probably be slightly sticky and pink (the edge of the tail is usually sticky too). If she is doing at least two of these things, you may consider her in heat. Either take her to a male, making an appointment beforehand; or you may wish to check your findings by leaving her until her next heat, which will be in 21 days if you are right.

As there is a definite rutting season for goats, from September until about mid-February, it must be during this time that your goat is mated. Once having begun her cycle, she will 'call' every 21 days for the greater part of those five months. Calling means literally that in many cases, particularly the white breeds and some Anglo Nubians and Alpines who have a name for being vocal.

Always it seems to be the one who is 'running through' (that is, not being served until next year) as her milk will be needed in the winter, who is the musical type.

One really useful attribute of the goat is this ability to milk for two years between kiddings, so that two milkers can supply a household without any dry period if sensibly managed.

AFTER MATING

Once mated, the goat should be watched at 21 days, and again at 42 days for a return heat; after that, she is presumed in kid. A goatling will from about ten weeks on begin to assume a more solid appearance, and her rations should gradually be increased. She will need more food to keep herself and her unborn kids growing. By the time she is due to kid, she should be receiving as much as you expect her to need as a milker. The last week before kidding, add more bran than usual to her mixture; the week after kidding, bran and oats alone in fair quantity will be enough. No protein should be given, as the protein will push up the milk too quickly and might cause milk fever. This applies to a goat of any age, not only goatlings.

If your goat is a milker, she will begin to go down in milk about eight weeks after service, and will in all probability dry herself off about the 14th week. Should she keep on too long, discourage her by milking at uneven times, milking once daily only and so on, but if this does not do the trick, don't worry. Feed her well, as some goats will never dry off altogether, or it might be a 'cloudburst'. A cloudburst to most people is merely a very heavy downpour, but to a goatkeeper it means a pseudo-pregnancy. The goat will appear to be in kid, expand visibly, but not dry off; and she will not soften round the tail, as an in-kid animal will. She may well go on longer than the 152 days of normal term gestation, then one day you will find a nice slim goat, with a very wet bed screaming for her 'waterbaby' as she will consider she has kidded, and is looking for her offspring. She will have the usual post-kidding discharge. This is something which worries many beginners in the goat world, as goats have a blood discharge from about a week after kidding for the following two to three weeks. This is normal and nothing to worry about; it is usually slight but continuous.

SIGNS OF PREGNANCY

About six weeks before kidding, there is a definite softening round the tail. In a non-pregnant animal, it will be solid and bony on the back beside the tail root. In the forward in-kid goat you will feel a soft patch both sides, skin, not bone. This is the bones loosening for the kid's exit. Also, in the younger animal the udder will begin to grow and continue to do so but remain soft until kidding is imminent, then fill rapidly and harden. Older goats who have kidded previously will not start so soon, but may well make a very large bag; this must be watched to see that it does not get hard and shiny. Should this happen, some milk must be extracted or she will develop mastitis, but only remove enough to soften—do not strip the udder. This will not deprive the kids of colostrum, as that is only produced by the act of birth. Colostrum is the first milk; thick yellow custardy in appearance, it is vitally necessary to the young. It supplies all the minerals, vitamins and antibodies and is also a laxative which clears from their system all the meconium—the material which has supplied their needs whilst in utero. The first excreta of kids is black, and once milk is being digested, turns yellow; thus you know all sections are working as they should.

Birth is natural, but nevertheless it is a good thing to keep your eye on the situation, since births in goats are multiple. So many long legs, where two or three kids are involved, can sometimes lead to a confused arrangement. A few days before the kidding is due, you should have prepared the pen where your goat is to kid by a good clean out and scrub down with disinfectant; the floor should be swilled and when dry, a good thick bed of straw should be laid down, you should also have a nice deep box, wooden tea chest type, in which your kids can be kept warm. Place a layer of straw in the box, and collect some old towels or kitchen roll, soap, nail brush, dettol, iodine, and put the whole lot in a polythene bag to keep clean and dry in the box. You have done all you can to prepare, it's up to the goat now.

ARTIFICIAL INSEMINATION

Artificial insemination of goats is still not widely used, but a growing method. East Anglia was the first area where a breeder was licensed to abstract semen and prepare straws for trial use from specially designated males of the highest breeding. But since the Caprine and Ovine Breeding Society (COBS) was formed, there has been a great advance. Many countries—among them Denmark, Switzerland, France, the former Soviet Union, Japan, and the USA—have schemes working, but once more the rugged 'individualism' of the British goatkeeper makes for a very slow take-off. The benefits from using AI are so convincing that it seems unbelievable anyone should keep a male goat when they could, for the sake of a telephone call, take their pick of the 'best' the country could offer.

The main argument against AI is the cost. Naturally it is not cheap, as storage and various other costs are high. Nevertheless, costed against transport to the male of your choice, the stud fee, and your own time (probably the most costly item) AI looks reasonable. And should the animal return to service, a further insemination is available—there is no need for another dash across the country, usually on a dark and horrid evening!

To export animals, both the carriage cost and time involved are very considerable, whereas semen could be flown to any destination, where the local laws permit, in a matter of hours at a vastly lower cost.

Newly imported straws from Switzerland extend the selection of Toggenburg sires considerably. A list of officially trained goat inseminators is available from the British Goat Society.

SELLING STOCK

It is essential to inspect kids carefully, whether for your own herd or for sale. Make sure there are no extra fishtail or double teats. Should any be found, the animals must be culled as they may reproduce this fault.

Never repeat a mating which produces faults of any kind. They will probably not happen if a male from a different line is used. Only a really sound beast should be used for breeding.

Rearing is very expensive, so the age at which you sell is important. Kids which are being bottle-fed can be sold from one week; this obviously gives the largest margin of profit. Milk taken up to that time is small and mainly colostrum, useless for other purposes anyway.

Many breeders, however, like to sell stock as first kidders. They can then try for a Star in the milking competition at an agricultural show. This gives a certain amount of prestige to a family, and the longer it goes on unbroken the better it is thought of. If you buy such a young milker, she will cost quite a lot of money; likewise, when selling such a goat, you must ask a realistic figure as you have kept her for two years plus. The goatling is easy to look at; potential buyers can see all her good and bad points, have the whole of her lactating life ahead of them and, apart from udder shape, know the best and the worst. As the expensive part of her rearing has been done, she should be a fair price, and she should not cost much to keep for the remaining time until she is served.

Breeders generally do not advertise in newspapers, as good animals of any age are in demand and many are booked before birth, or spoken for as soon as it is known that they are for sale.

Sometimes, breeders advertise in the British Goat Society's monthly journal, which gives more folk a chance to inquire about the stock for sale, but this is very much 'preaching to the converted'.

Males are somewhat different; very few are sold compared to females, which is understandable. One male is capable of giving many services in one season alone, and as only the best progeny of the best milkers is kept, many male kids find their way as carcases into the catering trade or the household deep freezer. Well-established herds, with some show wins and milk records to show to purchasers, are likely to have their males considered worth buying.

Some breeds are more popular for exporting than others, and only high quality stock is bought for this purpose. Orders usually come through the British Goat Society. Suitable animals are always wanted within well-defined limits and are never younger than weaned kids, as milk for rearing is short in many countries, especially hot areas. Every country has its own set of tests to be carried out by a veterinary surgeon. When the correct form is received, the animal must be examined by a person representing the BGS, and all animals must be tattooed in the right ear with a number and letters issued by the BGS. Those who think that at some time they might wish to export must register now with MAFF.

All males must be so earmarked before they can be registered with the Society.

New regulations via the EC state other requirements, so be sure to get the latest information on this before attempting to sell anywhere at all as some areas may be affected by MAFF 'stand still' orders because of some infection or other. The British Goat Society will give up-to-date views. Possibly your local vet will also know what is allowable.

CHAPTER 6

KIDDING

WHEN a goat is ready to kid, normally her udder will fill up very rapidly, her flanks will hollow, and the tail bone on top of the back will ridge up and feel very slack. She will 'talk' to her sides and become restless, getting up, snatching hay, laying down again and so on. She may refuse her food. Remove her water pail to prevent accidents to kids. She will most probably paw up her bedding, make hollows in it, and generally get more distressed. A small amount of colourless discharge will appear from the vulva. All this may take anything up to twelve hours, though some come along much more quickly.

Eventually she will begin to strain with greater frequency and strength. A membrane bag will appear from the vagina and in it should be two hooves, one slightly before the other, then the tip of a nose should be seen; this is all as it should be.

Straining now will usually be accompanied by a cry, and some goats are very vocal, but this does not mean anything is wrong. A few more good heaves and the head and front legs will be thrust out, and very soon the remainder of the kid will slide out, which is the time you should appear. Wipe its nose and mouth with a corner of a towel, removing any mucus, going round the tongue to make sure the kid can breathe; place it near its mother's head and she will begin to lick it clean and dry.

Keep a watch on the dam, as she may well strain again very quickly, and you would do well to pick up the kid and put iodine on the cord which is raw and hanging under its body; this will prevent infection getting in, and also help to dry up the cord quickly. Continue to dry the kid, then put it into an upturned box, so its mother will not stand or roll on it whilst producing the next kid. The process is then repeated.

BREECH BIRTH

With three or four kids, you often get one of them coming as a breech, which is tail first. You wonder what on earth you can see, and if this is the case, you will have to scrub your hand well and soap your fingers or use slip gel. Holding your fingers in a cone shape, insert them gently into the birth channel at one side and bring down one leg. After rinsing and re-soaping the hand, do the same on the other side to bring out the second leg; this will allow the kid's body to come away. When the rib cage is emerging, hold both hands round it to stop expansion. If at this stage, the kid tries to breathe when the cold air stimulates it, it will gulp in fluid and may well be drowned before you can get it away from its dam. Once born, hold it head down for a few seconds to drain, then proceed as before. Whenever you enter the animal, you always take in infection. Therefore always use antibiotic pessaries as a safe-guard.

If each kid is placed in the box and left there for a time, you can wipe the goat's udder with a warm damp cloth, then milk some colostrum into a previously warmed empty wine bottle and make sure each kid has a few ounces. This will warm them and keep them quiet and content until their dam has 'cleansed', that is, dropped the afterbirth. Usually it takes about two hours, and should come away cleanly with no pieces left hanging.

Should she not cleanse, but still have the membrane hanging, don't on any account pull it as that can cause a haemorrhage. But if after twelve hours, she has still not got rid of it, you must get veterinary help quickly. A goat closes up rapidly and cannot be left like a cow. She will need some hormones to make her contract and drop this membrane. Usually the vet will also give an antibiotic injection at the same time, in case of infection, but thank goodness this is most exceptional and very seldom happens.

REMOVE THE AFTERBIRTH

So our nice, normal animal has dropped her afterbirth, and you have removed it to prevent her eating it. In the wild, it would be normal for her to do so, removing evidence of young and helpless creatures from predators. However, the domesticated goat does not have the digestive abilities of her wild kin, and would get bad indigestion if she was to eat all that rubbish. Give your goat a warm drink. Some people use oatmeal water, but warm water with a tablespoonful of salt will be appreciated; she has a large gap amidships to fill after several kids are removed, and warmth is a good thing.

At this stage the kids' box should be placed in the corner of mother's pen, turned onto its side (kennel fashion) and the kids allowed to stagger around their dam, who will continue to lick and tidy them up. Her first meal should be a warm bran mash, of about one pound (453 g) of bran and a handful of salt, just dampened with hot water.

You must see that all the kids can suck, and having done that, leave your new family to have a rest. Go and have a cup of tea; you will find you are almost as tired as the goat to begin with! I knew a shepherd who used to take whisky and lambing drink in his pocket, one for each of them, and he didn't have the lambing drink.

Kids are generally left for four days before being removed from their mother's care, and then are bottle fed. Thus, you can find out how much the kids are taking, and dispose of any kids you do not wish to rear. After four days the milk will be normal; it will not be of any use, except for the kids, until then.

It is a good idea to keep some colostrum. There is usually much more than is needed by the kids, and any excess can be placed in a plastic container and frozen. This can be used in case a milker is unable to supply colostrum for her kids: occasionally udders do not fill up until after kidding, or a bad kidding may exhaust the dam so that the kids are unable to get to her to suckle. Just warm it; do not overheat or it will become thick like junket or egg custard.

About a week after kidding a worrying blood discharge may appear from milkers. This discharge is quite normal and may go on for anything up to three weeks. To avoid any discomfort simply wash daily with warm water which will prevent it hardening and avoid loss of hair from the tail.

MALPRESENTATIONS

There are a number of malpresentations which can occur, and which are not beyond the competence of a reasonably level headed person to deal with. Firstly, if just one hoof can be seen and no nose, scrub your hand, and if you have someone to help you, get them to hold the goat's head so she cannot move off.

Insert your hand as suggested previously, but push the foot back into the uterus. You will probably find that the other foot has doubled under, and once you have unfolded it and made sure the head is atop the legs, all should go ahead.

If the head is not visible, again push the legs back and insert your scrubbed hand and arm into the uterus; put the head into the correct place; it may have been pushed up and caught on the bone above the uterine opening, so that pushing it back will allow it to fall into position. Or it may be turned back or sideways; but once you have corrected the misplacement the kid can arrive properly.

Once in a while you get two feet, which are not from the same animal. As before, the whole lot must be pushed back gently but firmly; then find which kid is closest to the cervix and push the second one back out of the way. You will be surprised how much space there is in which to manoeuvre them around, but after an assisted birth it is sensible to have the goat given an antibiotic in case infection has been carried in.

THE GOATLING

Kids would normally be mated in the first year of their lives, and indeed in Europe this is the case. However, at one time in the British Isles some of the animals being bred were considered to be too small for the purpose of producing good strong and bigger offspring, so it was decreed that a store year, which is known as the goatling year, should be created.

The young animal is therefore about 16 to 20 months old (according to the month in which it was born) before she is served for the first time, so is of a greater size and well able to carry the greater milk load which it is hoped she will produce.

Being the somewhat contrary animals that they are, some goatlings do produce milk at this stage and are known as 'maiden milkers'. Some will give several pounds of milk daily from spring until they dry up after being served in the autumn. The udder does not always develop evenly, but this is not to say that it will not do when the animal kids.

This larger first kidder is more able to stand up to the two-year, or extended, lactation which we expect from our animals in the United Kingdom.

As goatlings have no particular acts expected of them, they just eat, play around and generally misbehave. They are the delinquents of the goat world, but nice with it!

CHAPTER 7

REARING

HAVING given plenty of thought to the various commitments that go with kid rearing, such as being tied to regular feeding times for several months, and considered the amount of milk or the cost of buying milk supplement for the kids, do you really need a herd replacement or additional stock? Then, if you do decide to rear one or more kids, make absolutely sure that the particular kids are worth the time and money that will be expended.

Examine each one carefully for any sign of abnormality and keep kids only from the best milk line. Select the kids you personally like the best, otherwise they may become too tiresome to keep going and the project be abandoned after a time.

If it is possible to keep more than one kid, do so; they like company and you can always sell one at a later date.

Kids are very active. They like to run, jump and generally leap about, so they should never be tethered, even if the older animals must be restrained in this way. If they have no yard, it is better to keep them in their house and take them for walks.

Get your kids to come to you when you call. Train them to walk on a lead—you may wish to show them or take them browsing along a lane. Don't allow them to jump up against you; they get heavy and also soil clothing. My own goats are trained to answer to hand-clapping; I attract their attention if at a distance in this way and can also steer them in the right direction by clapping to the right or left, as required.

FEEDING METHODS

There are several schools of thought on feeding:

Natural rearing. In other words, leaving the kid or kids on their dam to suckle. Of course there is much to be said for this, especially if one is at work during the day, or occupied in such a manner as to make it impossible to feed kids at regular intervals. Certainly the kids get all the milk they need, very likely much more, but they do become much harder to handle, being unused to human contact.

Against the good must be set the fact that not all goats like their young around them all the time, and would prefer to graze and cud in peace. Also, you still have to strip out the goat at morning and evening, or she will get overstocked on one side as kids almost always feed from one side of the udder only, which means one side is distended, whilst the other never fills at all, so the result is an uneven udder.

It is quite a good idea to leave the kids with the mother during the day, remove them for the night and strip her out. In the morning, give the kids a bottle, milk out the goat and let them all out together. In this way you get all the overnight milk for the household, and the kids get their share during the day. The morning feed can be dried milk if there is not sufficient to spare from their dam after the house milk has been collected. A box placed on its side kennel-wise is loved by all kids, and between feeds very young animals will curl up in it and sleep; later on great fun is had jumping on and off the box. If the weather is fine and dry, kids can run in and out for short spells when the older goats are in. In this way they will not be tempted to try and suck from mother. It takes about three weeks for them to forget and treat their bottle and its handler as mum.

Kids left to suckle should not go on to the grass until they have eaten some hay, otherwise they will probably fill up too much on the lush grass and then scour. If this means bringing the milker in at lunchtime, so be it. On the other hand, if you are out all day it would be better to keep the milker in for two or three weeks until the kid is eating a little, after which the kid will be all right. Suckled kids take smaller feeds more often, so do not allow them either a stomach full of milk followed by a load of grass or the grass followed by milk because this causes scouring and, possibly, bloat.

Should a kid scour, it may be caused by the milk being fed too hot or too cold, or by a dirty bottle. Reduce the next feed, and all subsequent feeds for 24 hours, to half strength and add a heaped teaspoonful of light kaolin powder to the milk at each feed until the droppings become more solid. Then gradually increase the strength of the milk and leave out the kaolin.

Some of the author's very young kids still damp from birth and being bottle fed.

TAKING TO THE BOTTLE

Naturally a kid does not recognise the alien object, the teat, as anything of use or pleasure. Consequently, to teach a kid to take a bottle may take a little time and patience. Some are much quicker than others, but if you back the kid into a corner, stand with your back to the kid and bring its head between your knees with your left hand held below its chin, open its mouth with your thumb and forefinger, and with your right hand insert the teat into its mouth, holding it there with the left hand, it will soon learn to suck. If it cannot lower its bottom jaw, it has to keep the teat in its mouth. Kids who are given a feed of colostrum from a bottle very shortly after birth, whilst still drying off, are never any trouble to bottle feed later on.

Male kids are extremely precocious and can give effective service from as young as ten weeks; so it will be seen that a mixed pen of kids is a bad policy and can lead to accidents. Separating them at the start may be difficult if they are kept for company, no other female kid being of the correct age group. Only males from really well bred parents and good milking dams should be kept as it costs the same to rear a poor quality animal as a first class one. The other little lads can be used for tomorrow's lunch. If males are being reared for meat, they should be castrated by the ring method at a week old, at the same time as disbudding and earmarking.

Male kids for stud need very good feeding, as they begin their working lives from about six months. They should have one bottle a day for a month longer than their sisters, and a slightly larger concentrate ration. Sometimes at the beginning of the rutting season male kids lose their appetite, so a slight build-up in condition prior to this will help over the lean time.

Suggestions for feeding are as follows:

Four days to one month: all the milk kids will accept up to one pint (0.57 litres), four times daily.

One month to four months: one full bottle, three times daily.

Four to six months: one bottle morning and evening.

Six to seven or eight months: one bottle in the evening. Concentrates should be fed at the same time as the other goats are fed, a.m. and p.m.

Bowl feeding. Kids, after having been left with their dams for the four days during which they take the colostrum they so definitely need, can be taught to drink from a bowl, pail, measure or any article of a suitable capacity. The kid's nose must be held just touching the milk; it will lap to begin with, then drink. This method does work, but it is not really recommended for the following reasons.

It is natural for a kid to suck; in doing so the digestive juice gets mixed with the milk and also the milk by-passes the rumen and goes direct to the second stomach (masum) where it is digested correctly. By drinking, on the other hand, the milk goes into the rumen, is gulped down very fast, and does not digest so well, air is taken in with it, and the kid gets various ill effects. Also kids, so fed, tend to get pot-bellied, which is a far from a pretty sight.

Bottle feeding. This is the most favoured form of feeding. Kids are again left for the crucial four days to have the colostrum or, if removed immediately for any reason before then, are fed several small feeds of colostrum during the day milked directly from the dam into a bottle.

From four days on, feed four bottles of milk daily at 103°F (40°C), just over blood heat, of whatever quantity (up to one pint/0.57 litres) the kid will take. Once a whole pint (0.57 litres) is taken eagerly—at about one month—the kid may be fed at longer intervals, three times daily, and given a full bottle. Bottles used are usually the wine or mineral water type with a flat base to ease cleaning; the teats may be plugged in or pulled on. The type used for lambs is satisfactory and can be obtained from a good dairy supplier or veterinary chemist. The British Goat Society also holds a stock for its members and will always help in an emergency.

Special milk feeding apparatus for self-help feeding by kids is now available from Volac. This firm also makes a new milk feed for kids which can be mixed, and fed, cold which is said to be better than offering the milk warm and allowing it to cool later. For those out at work this is an invaluable help, especially as many people are now rearing neutered male kids for meat; rising prices make it foolish to kill and bury kids which could be used to replace other expensive meats and assist the family budget.

For kids being reared for meat, younger ages for weaning are common. Eight to twelve weeks is considered adequate and much more economical as milk is such an expensive commodity. Rearer pellets can be used for feeding, and they are especially popular in France where most of the milk is used for cheesemaking.

HAY IS ESSENTIAL

Milk is not the only food required, but it is the first. Hay is essential for the cud to begin. From a very few days kids will nibble hay, which should be placed in a rack or a box so that they can reach it. If it is hung in a small bundle, be very sure that the loop is not large enough for the kid to get its head into, or the rails in the rack so far apart that the same thing happens. Kids panic if they are held fast and can strangle themselves.

A very small amount of concentrates should be offered from about two weeks. Place in a small bowl (a washing-up bowl with a brick in it to prevent it tipping and so wasting food is sat isfactory). Feed this small amount until they find they like it; then increase it gradually as the kid grows, giving about ½ lb (225 g) in two feeds.

If milk is needed for the household, you can either mix what milk can be spared with a good calf-milk substitute or you can use the latter entirely after one month. But try and give whole milk for the first month. There are many excellent powder milks for calves, in fact most manufacturers have one in their range of dairy products. Orphan lamb foods are also good but are too strong for the kids so use at half strength. Kids have milk for so much longer than lambs who have to assimilate their milk quickly, or not at all, as ewes become dry quickly.

VACCINATE AGAINST ENTEROTOXAEMIA

The worming of kids which are grazing is essential (see page 106). I have no particular liking for introducing unnecessary medications into any animal. The one vaccination which I do insist on having is against enterotoxaemia. This is given into the loose skin behind the neck and is simply an injection of 2 ml of pulpy kidney and tetanus vaccine. At one month a kid receives its first dose. All kids need a second vaccination one month later (see page 106), and after that, once every six months, though in some parts of the Southwest particularly where too much lush grass causes problems, breeders have found it necessary to vaccinate at four monthly intervals. Several kinds are being marketed, so vary the vaccine you use. Sadly, most remedies are sold in quantities far too large for the average goatkeeper, but I have found it practical to share with others, injecting my own goats and then those of other folk in the area, or giving them the containers so they could do their own goats when convenient. The next time someone else did the buying and started the round of vaccinating. This way the material was used up whilst fresh.

Most of these vaccines contain an anti-tetanus vaccine, which is excellent in preventing infection from thorn pricks or sore patches caused by head banging (which all goats do as entertainment at times). Cheap disposable syringes are easily bought and simpler to use than the glass type. If a sterile needle is pushed through the rubber cap on the flask and left in position, all syringes can be filled through it, and a different needle used for each animal. It is possible to sterilise needles in boiling water. Store them in surgical spirit in a closed container.

It is definitely not worth risking your animal for the cost of an injection. Should enterotoxaemia be encountered, it will then be a mild form which can be cured by giving an antibiotic injection and, orally, sulphadimadine (5 g for each 50 kg of the animal's body weight). It is not possible to save untreated animals—the action of ET is so rapid.

DISBUDDING

The law decrees that no unnecessary suffering may be inflicted on animals, so the process of disbudding—that is removing the growing horn which can be felt below the skin of newly born kids—was changed. It used to be possible to disbud your own stock and that of others if you could do the job well. Most Goat Clubs had one capable person who would do the disbudding for all who asked, in return for a small contribution to the cost of the gas used to heat the iron. Kids must now be done by a veterinary surgeon who uses an anaesthetic, which is unfortunate as it causes much more trouble than the actual removal of the horn. But such is the case and you must abide by the law.

When the hair is dry (so that it is tight over the horn bud and doesn't slip), run your fingers over the top of the head. Should you feel a bump over which the skin feels tight, you have a horned kid and should immediately make an appointment with your veterinary surgeon to have the horns removed. This should be done before the kids are a week old.

Keep the top of the head dry once disbudded. Milk from the bottle or any rain spots should be wiped off or the small scab that forms will get messy. It should dry up and fall off after about a month. Just a small hairless patch will be seen which will eventually be covered by hair. Run your finger around the area and if any small points of horn are felt arrange with the veterinary surgeon to sear them off. Otherwise an ugly piece of horn will grow, spoiling the animal's appearance; it can also get knocked off and cause a mess if nothing worse.

Do not be persuaded to leave the horns on because the kid is being reared for meat only. Horns can injure without the animal's intention, for example, by bringing a head up under a feeding pail or catching a horn on the back of your leg.

EAR MARKING

Kids also have to be ear marked before registration to allow identification on inspection. The mark must be placed in the right ear. The BGS issues the letter annually, missing out 'O' etc which could be misread. Clubs will give a number to a member, and those with several animals can ask the BGS for a personal prefix letter or letters, which may then go on from one year to the next, e.g. AZ24A, AZ25B, though most people begin again from one each year, e.g. AZ1C etc. Clubs have a special set of letters and they allocate a number for each kid as you ask for them. You can request a personal number from the BGS if you are likely to have many kids to register. When showing, this number has to be included on entry forms. When a female is served, the registration number has to be placed on the Stud Certificate and of course the male's number is also included. Should you hope to export stock, a special number must be requested and is placed in the left ear.

Remember that kids cannot be registered until they are a month old, and that all the kids from one litter must be registered together. The owner must belong to the BGS or one of the many goat clubs recognised by the Society. A prefix can be registered for a fee. The club secretary will advise you on procedure and cost, also on other useful details.

CHAPTER 8

MILKING

MILKING is not difficult, but to begin with it does make the hands ache a little. Firstly collect together all you will need—a milking pail preferably stainless steel, a container topped by a strainer, a milking stand, raised so that you can see what you are doing, which could be made of wood and covered with polythene for easy cleaning or concrete which can be hosed down and, of course, a goat.

Place the milking stand in position, or stand the goat at an angle. If you feel unsure of the goat, fasten her up with a collar on a short cord attached to the wall. Since hygiene is of the utmost importance, it is a good idea to place a cut-down filter bowl with a new filter paper for each milking in the milking pail to prevent any hair or foreign bodies from falling into the milk. Either sit on the edge of the stand, or on a stool beside it, and then:

1. Use an udder wipe for each animal and rub round the goat's udder to be sure it is clean. This will induce milk 'let-down'.

2. Place one hand on each side of the udder, with the teat lying across your fingers, and your thumb on the outside of the teat. Cut off the milk from the udder by compressing the top of the teat with your thumb; then squeeze the milk down towards the pail by closing your fingers one after the other until the teat is empty, then relax your thumb, which will allow more milk to come from the udder into the teat. Do one side, then the other alternatively until the udder appears to be empty.

3. Then, rub both hands round the udder from the top where it joins the body down to the teats; this will bring down more milk. Remove that and do the same again several times until all the milk is extracted. The last part milked contains the butterfat; if you do not 'strip' her out, she will give that much less next time.

STRAIN THE MILK

When you have finished milking pour the milk through the strainer into the vessel which is to be used for collecting, and go on to the next goat; or if only one is being milked, place the container immediately into cold water. Cool the strained milk immediately as unwanted and potentially dangerous bacteria double every twenty minutes at 12°C. The goat can then be returned to her pen.

So long as the container is fitted with a good lid, it is immaterial whether it is metal, stainless steel, china or plastic, providing it is easy to wash and sterilise. Strainers are also expensive pieces of equipment. Milk strainers are metal and bowl shaped. They are narrower towards the neck which contains two metal discs pierced with round holes; these are loose and the milk filter is put between them and the whole lot put into place. The milk pours through this into the churn below. Two nylon sieves with a milk filter trapped between the sieves will prove satisfactory and a cheap substitute. Place across the open top of your churn. No foreign bodies should have found their way into the milk.

Dairy detergent and Steralent are available from a dairy sundries supplier. It is called hypochlorite.

DAIRY EQUIPMENT

Your milking pail should be stainless steel. All equipment used for milk should be rinsed with cold water, washed with hot water and detergent, and then rinsed out with steriliser, which is left in for at least thirty seconds. This gets behind bacteria which if left, causes off flavours and souring of the milk. Finally, rinse with clear water; a special churn brush is well worth buying as it gets into the most inaccessible places; it is inexpensive.

If you are selling produce (or giving it away) for human consumption you are deemed to be commercial. Your premises will need to be registered with the local Environmental Health Office as from 1992. You will also have to attend and pass a food hygiene training course.

Milking should always be done outside the goat's normal pen, preferably also away from hay and dusty straw. The area must be washed down after milking to keep the whole place fresh and clean.

Dairy equipment – milking churns (1, 2 and 3 gallon/4.5, 9.1 and
13.6 litre), pail, strainer and churn brush.

COOLING PROCEDURE

Cooling of the milk should be done immediately after milking is
completed. As has been said, the equipment can be very basic as
long as it cools rapidly. Stand the container in a bowl in the sink
with the cold tap trickling over it, or have a tank with a tap at
the bottom and an open top. Stand your churn in the tank, place
a hosepipe on to the water tap and let the water flow in slowly.
Open the tap at the bottom to allow a flow of cold water to go
round, until you consider it is sufficiently cool, 5°C (40°F). Turn
off both taps and leave the churn standing in the cold water. In
summer tap water can rise in temperature so continue to run the
water, or refrigerate the milk, until it is cooled to below 12°C.

A sparge ring is another method. It consists of a piece of hose
piping, large enough to surround the neck of the churn, and is
pierced with holes. Obtain a tee-junction, fix the hose piping on
to each end; the third part of the tee is fitted to a length of hose
which is attached to the cold water tap; when turned on, water

trickles down the sides of the churn. There are in-churn coolers, which are mostly for 8 gal (36 litre) churns and not many goat-keepers have that amount of milk at one milking. The same goes for ice-bank coolers. Alternatively, an old fridge in the milking area can be used.

There are also in-churn coolers to fit much smaller churns, some for 1¾–2¼ gal (8–10 litres). Plastic tubing swirls down into the milk, carrying cold water, which rises to force its way out of a perforated-edged lid and finally runs down the outside of the churn into a sink or tank. This very efficient cooler is priced at around £50, available from Lincolnshire Smallholdings Supplies. I have also obtained churns from them made of Dairythene and holding 1¾–2¼ gal (8–10 litres) of milk, which are far lighter than metal ones and not so highly priced either.

Goat's milk is very palatable. It does not have any odd strong flavours as is so often said; in fact, off flavours reflect badly on your dairy hygiene.

MILK PRODUCTS

Cream is light in texture and delicious; it can be obtained by separating or clotting.

Butter can be made from either separated or clotted cream, but unless you like a white butter, some anatto for colouring is necessary.

Cheese, both soft cream and hard-pressed varieties, is simple to make if directions are followed and the correct cheese rennet is bought, again from dairy suppliers. A lactic starter can be used, obtainable from the same source.

Yoghourt, so very popular, is simple to make, with or without the range of electrical gadgets advertised for the purpose.

You can use a vacuum flask for small amounts or make a do it-yourself 'warm box', which works very well. All you need is a wooden or tin box, deep enough to take a basin with a lid or a large jar. Pack the box—all round the basin or jar—with vermiculite or expanded polystyrene. Whilst the yoghourt is being made, fill the jar with hot water. When ready remove the hot water, pour in the cultured milk and replace the jar in the box; then cover with more polystyrene and replace the box lid. Leave the jar overnight in the box, then remove the yoghourt and cool.

Curds, sweetened and with added fruit (used for Yorkshire curd tart or American cheese cake), can be made with little trouble or expense.

Owing to the small fat globule and the vitamin A (with little casein) in goat milk, the milk freezes well depending on the speed of initial freezing and thaws correctly, if thawed slowly overnight. This is most useful as summer surplus can be kept for use in winter when the yield goes down somewhat.

Do not be tempted to take on customers in the summer, to use the extra milk, unless you feel sure that you will have sufficient milk during the winter to continue the supply. Anything to be sold must be labelled, which is a real can of worms. Please do not go into this without first getting as many leaflets from the Ministry of Agriculture, Fisheries and Food as you can assemble. They are complicated, but the information is all there somewhere. Goat produce comes under 'Dairy produce' not under 'Milk Regulations'—it appears that only cows are allowed to produce milk officially. Containers, i.e. bottles, bags, cartons, boxes, yoghourt tubs and cheese packs, are supplied specially for goat produce and have the fact that they are goat produce printed on them. The Trading Standards Office is most helpful with labelling. Do be quite sure your name and address is printed in really large letters either on an adhesive label or on the container itself. The contents must be listed in descending order of quantity.

Produce must be labelled with 'Use by' or 'Best before' dates. 'Sell by' is no longer permissible. Raw goats' milk and products made from raw goats' milk must be labelled 'unpasteurised'. Check with the Trading Standards Office.

You need to take a short course in food hygiene and obtain a certificate, which you can wave at the authorities if required to do so. Many colleges run courses for this purpose. Life gets more complicated with every new EC rule and regulation. Contact your local Environmental Health Officer for further information. The EHO will test your milk occasionally without prior notice to ensure standards are being maintained.

KID MEAT

Kids surplus to herd replacements or males not good enough to keep for stud, if reared for a few weeks, make very tasty meals. Very young kid, say 4–6 weeks, is like chicken; to four months, like veal; over that, like lamb. But kids being kept for meat should be castrated by the ring method before one week old, to prevent accidents and stolen matings later on.

MILK RECORDING

Milk recording is carried out by members of the British Goat Society and its affiliated clubs or societies. Each month, milking to a strict timetable, members weigh each goat's milk and take samples for analysis of its butterfat and protein content using measuring rods to obtain an accurate sample of milk. You will need an approved weigh balance. The measuring rods, sample bottles and all paperwork will be provided by the Milk Recording Club Secretary.

In the evening each goat is milked, the milk weighed and the yield recorded on a parlour sheet. The hollow measuring rod is dipped into the milk and the milk in the rod is then placed in the sample bottle for that particular goat, and so on until all goats have been milked and samples taken. The following morning the routine is repeated. The second milk sample is placed in the appropriate bottle containing the previous sample taken the night before, the cap fixed on tightly, and the bottle shaken well. The evening milk is almost always higher in butterfat than that of the morning, hence the need for two samples mixed together. The protein content from evening and morning milking tends to be similar.

The sample bottles should then be wrapped in clingfilm to prevent leakage and, along with the parlour sheet, should be posted to the Milk Recording Club Secretary as soon as possible after morning milking. The milk samples are sent to a laboratory for analysis; then the results, together with the figures for the yield and other necessary information, are sent to the British Goat Society Secretary. A copy of the monthly record for the herd is returned to the member. At the end of a goat's lactation, a certificate indicating the total yield, number of days in milk, and percentage of butterfat and protein is sent to both the member and the British Goat Society Secretary. Annually all yields which have been achieved by means of at least six butterfat/protein tests are published in the British Goat Society Herd Book, subject to the approval of the British Goat Society Committee.

To standardise procedures and to create confidence in the records, Club Recording Officers, or people nominated by them, visit herds, without notice, to make check recordings.

Providing that the goat has given at least 1000 kg with an average of 3% butterfat over the lactation, she will receive letters and figures to be put before her name. For example, if she gave 1000 kg in a period of 365 days or less, she would qualify for the prefix R100. For each additional 10 kg she would be allowed another 1. Thus 1010 kg would qualify for R101. All yields are now recorded and displayed in kilograms.

The British Goat Society publishes a very useful Milk Recording Manual which goes into great detail concerning all aspects of goat milk recording. Copies are available at a very reasonable cost from the British Goat Society Secretary.

The purpose of milk recording is to obtain reliable and standardised information about the yield and the quality of the milk produced by each goat in a herd. Milk records can be used as an aid to improve management, assess stock for breeding purposes, select progeny and qualify a goat for a milk recording award. However, milk recording can be used for much more than a means of qualifying a goat for an award. Milk recording builds up a long term record of herd performance against which can be measured the effect of changes in management and stock improvement. Because of this every milker in the herd is important. Even if none of the goats in the herd qualifies for an award, milk records provide a sound basis for herd improvement.

DISTINGUISHING MARKS

Other distinguishing marks are AR (Advanced Register) and RM (Register of Merit) for high yielders and their daughters. Males from dams gaining their R, and whose sires' dams have also got this award, are indicated by being prefixed with a 'SS'. This is followed by the figures for the dam (RM 182) and the sire's dam (RM 172), thus SS182/172. If both the females have also won milking stars, he gets SS182/172,†—the latter mark being a dagger. This makes it easy for you to see at a glance which herd sires are from recorded herds, and the sources from which you are likely to get a male capable of bringing more milk into the herd. It is most interesting to note that often animals who do not have a very heavy yield in summer, but whose yield drops very little in autumn and winter, produce a very good overall quantity. The consistent milker is, in fact, the most useful to have.

MINISTRY TESTING

All goatkeepers must supply names and addresses to the local MAFF office and be prepared for visits to ensure premises are up

Highest Recorded Goat in each Breed (taken from the latest Herd Book 1989–90)

	1989–1990	Kidder	Quantity (kg)	Average BF (%)	Protein (%)
BS: Warkleigh Mitsoukou	BS029857ᴰ	1st	2617	3.59	2.61
BT: Neathwood Joclyn Q*	BT016277ᴰ	2nd	2131	3.40	2.66
T: Ch Pippins Cilla Q* BrCh	T00355Bᴰ	2nd	1948	4.32	2.78
AN: RM191 Norbury Amber Q*1 BrCh	AN18238ᴰ	3rd	1910	4.86	4.19
BA: R189 Carrick Rebecca Q*	BA012381ᴾ	3rd	1968	4.12	3.33
S: Prastens Penelope Q*	S005882ᴾ	2nd	1697	3.56	3.00
GG: R144 Telgar Blondie *	GG001776ᴰ	2nd	1443	3.29	3.01
HB: ChRM320 Ashdene Metaphor Q*2	HB051045ᴾ	4th	2965	3.40	2.63

(D = disbudded, P = polled)

CH R194 Pippins Cilla Q* Br Ch T35556D owned by Mrs D M Hearn, St Austell, Cornwall and bred by Mrs M Robinson. She has given the heaviest milk yield of any known Toggenburg.

to standards. Ministry officials now call at intervals on goatkeepers who sell milk. They take two samples for testing, one from the morning milk and one from the freezer. From these they are able to advise the milk's rating. If you maintain good standards there is no reason to worry, and obviously it is to your advantage to know if anything can be improved.

CHAPTER 9

DAIRY PRODUCTS AND BY-PRODUCTS

PRODUCE must relate to new regulations and testing for CAE, and premises must be 'correct for the purpose of production', so please check to find out what is necessary before spending time and money on alterations.

EQUIPMENT REQUIRED

A thermometer is really essential for good results. There is one specially made for dairying; printed on it are the degrees Centigrade or Fahrenheit for each purpose: 17°C (62°F) churn, 18°C (65°F) separate, 32°C (90°F) cheese, 77°C (170°F) pasteurise. It is not expensive and makes a vast difference to your cheese.

Heatproof mats are also quite cheap and help to slow down the heating rate. Rapid warmth makes the curd much harder. Whether using electricity, gas or solid fuel, always use a heatproof mat beneath the bowl or pan containing milk.

Wide cream-setting bowls can be bought, as can glass butter churns, milk pails, butter pats and cheese draining mats. Enamelled, plastic or metal alloy 18 lb (8 kg) bowls are difficult to find but are most useful should you manage to locate them, so look around your local hardware shop.

Both cheese and butter will be white unless you add colouring; this is named anatto. There are two kinds: the butter variety colours the butterfats only, whilst cheese anatto colours the solids not-fat. It is vegetable in origin, no E numbers.

There are many firms making cheese moulds from plastic, but you can improvise with plastic piping (provided it is made from food grade quality) which comes in many sizes. Drilled with holes all round and cut into suitable lengths for the amount you want to make up, almost any size of cheese can be moulded. A block of wood, made to fit the open top and placed over the drained curd, is

Equipment for butter-making – churn, mould, butterpats and stamper.

required; this is a follower, and weights are placed on to it to press the cheese. Use cheese cloths or muslin to prevent the wood from coming into contact with the cheese.

Bakers' racks or stainless steel or plastic cheese racks can be used for your cheese moulds. Shelves from a disused refrigerator or cooker can be used too. Cloths and racks should be sterilised. When the moulds are put into position, place a bowl under the racks or shelves to catch the dripping whey. Polythene boxes, drilled for drainage, can be used.

If you haven't got a butter churn, use a rotary egg whisk or an electric mixer; it is much less trouble that way and quicker too.

Cheese rennet and lactic starter are needed; junket rennet is useless. Vegetarian cheese rennet, for example Hesterhaze, is available from health food shops and dairy suppliers. Rennet is an extract from the pancreas of calves so strict vegetarians will not use it.

There is a special tool for ladling out cheese curd. It is saucer shaped and pierced to allow whey to run off as the curd is lifted, but if you cannot obtain one, then a shallow saucer will do fairly well. A plastic saucer with holes drilled into it works very well.

When milk is plentiful, you can make cheese curds and place them, after straining, into a box or bag and freeze for use during the winter. Butter, cheese and yoghourt can all be made from milk thawed after freezing.

Cream

To make clotted cream place the milk to be used in a wide topped bowl and stand in a cool place undisturbed for 12 hours. Place the bowl on a mat over a low heat, and leave until the milk just crinkles, remove and let it stand for a further 12 hours, then skim.

Should you wish to keep some separated cream, do not just put it into a carton and freeze it or you will be very disappointed as it will granulate. You can, however, keep it beautifully if you whip it and add a small amount of icing sugar; it will be very light and fluffy, and as good as fresh.

To make butter from clotted cream is quite satisfactory, but you will need a fair quantity to be of much use. So store up your

Electric separator in pieces.

cream from several settings; this will allow the cream to 'ripen' a little and add flavour to the butter. It should of course, be kept in a refrigerator or a very cool place.

Butter

If you have a glass churn, collect enough cream to half fill it, then turn the handle until the cream begins to break and form into grains. Add about half a pint (quarter of a litre) of very cold water to the cream; this is called 'breaking water'. Colour and salt—1 oz (28 g) to 1 lb (450 g) of cream—should be added at this stage; churn this in. Line a colander or other large strainer with muslin and tip all the butter grains into it over a bowl or sink; then run cold water over the butter, to wash out all the buttermilk. If it is left in, the butter will go rancid. When the water runs clear, take the corners of the muslin and squeeze out as much water as possible; then place the butter on a clean board or table and beat out the remaining water with butter pats or wooden spoons; then shape the butter into blocks and place in a cool place to set.

Butter from separated cream is made in exactly the same way, and both are equally successful and very nice in flavour.

Yoghourt

Yoghourt is very simple to make; again the thermometer is all-important.

Milk should be put into a double pan or a bowl standing in a saucepan of water. It should then be heated very slowly to 170°F (59°C) sterilising heat and held at that temperature for 30 minutes. Remove the pan from the heat, place the bowl into cold water and cool rapidly to 120°F (49°C), then add the starter; this can be a half carton of natural yoghourt. Mix into the milk and pour into a previously warmed vessel (you can use the 'warm box' described above). Put on the lid and cover and keep warm overnight. A vacuum flask can be used instead; simply fill it with the cultured milk and leave it overnight. Remember to cover the cork stopper with polythene, or next time tea and coffee is put into the flask it will have added yoghourt bacteria.

Jelly Cheese Cake—Children's Special

Place one green jelly into half a pint (280 ml) of hot water and stir until melted. Beat 4 oz (112 g) of cheese curds together with 1 oz

(28 g) of sugar until smooth. Line a cake tin, or a loose-bottomed mould at least 2½ in. (80 mm) high, with a biscuit crumb mixture and place in the refrigerator to set.

When the jelly is cool and begins to thicken, add the curds and sugar and whip until thick. Take the biscuit case out of the refrigerator and cover with a layer of any favourite fruit (orange segments are very popular) and fill the mould to the top with jelly curds. When set, remove from the tin and trim with butter cream or sugar trimming. It is delicious.

Simple Soft Cheese

Place one gallon (4.5 litres) of milk into a large stainless steel bowl, place it on top of an asbestos mat and heat very slowly to 32°C (95°F); then remove the bowl to some place where it can remain undisturbed for at least two hours. Take your cheese rennet and a small cup; put ½ teaspoonful (3 ml) of rennet into the cup, add two tablespoonfuls (20 ml) cold water, and mix. Stir the milk in a clockwise direction, and whilst it is swirling, pour on the rennet and water anticlockwise; this will mix it without need for stirring. Leave to stand for at least two hours or until solid, moving the contents away from the edge of the bowl with the flat of a spoon. Place a fine muslin or similar material in a colander and ladle the curds on to it. When all are in, take the corners of the muslin and tie together. Leave to strain.

After an hour or so, hang the bundle over a bowl to continue to drip overnight. Next day open up the bundle and if wanted savoury, add salt; you can also add chopped herbs such as chives or parsley, mix well and it is ready to use.

The same curd unsalted can be used for making cheese cake or sweet curd, as follows.

Sweet Curd

Take 1 lb (450 k) strained curd, place in a large basin and add 1 oz (28 g) brown sugar, 1 oz (28 g) sultanas, ½ teaspoonful (3 g) mixed cake spice or nutmeg, and mix. This is used as a tart filling, mixed with a beaten egg.

Cheese Cake (Lemon)

Make a base of good short pastry, or one of biscuit crumbs mixed with melted margarine, 4 oz (112 g) of crumbs and the same amount of margarine, and line an 8 in (203 mm) cooking pan with the mixture. Prick pastry to prevent rising, cook in a hot oven for 10 minutes. Take 8 oz (225 g) of cream cheese, whip until light, add two eggs, one at a time, and beat well after each addition. Blend in 4 oz (112 g) of sugar, rind and juice of one lemon, and half a cup of thick cream, beat all until thick and fluffy; turn into the baked pie shell and cook in a moderate oven for 15 minutes until set. Remove and cool; chill for one hour before serving.

Other fruit can, of course, be used instead of lemon.

A Hard Cheese

This requires more milk than soft cheese, and takes time to ripen.

Take two gallons (9 kg) of milk and place it over a very slow heat in a wide vessel, bring to 31°C (88°F), then remove from the heat and stand the vessel where it will be undisturbed. Add ½ teaspoonful of rennet to two tablespoonfuls (20 ml) of cold water; do not stir in as it will leave stir marks in the curds. Cover the bowl and leave one to two hours until set firm. Take a long knife, and holding it in an upright position cut the curds in both directions into cubes. This will allow the whey to come up to the top; remove some of it and replace with boiling water poured into the side of the bowl.

With the flat of the hand mix the contents of the bowl; keep removing whey and add boiling water until the whole mass in the bowl is 40°C (105°F). Allow the curd to settle; this process is called scalding the curd. When the curd has all settled to the bottom of the bowl, pour off the whey, and if you have a mixer, place the curds into the mixer bowl, add a tablespoonful (10 g) of salt for each lb (450 g) of curd, and beat smooth.

Make ready a mould, which can be weighted, line it with cheese muslin, and ladle the curds into it. Leave without any follower for an hour. Place follower over curds and add weights. These can be from a set of scales, or anything from tins of beans to the car jack so long as they add weight gradually. Clean house bricks wrapped in polythene bags are quite useful.

After two hours, turn the mould over and replace the weights on the opposite side. Turn and add more weight several more times during the next 24 hours. After that just turn night and morning for four days, using a dry mat each time you turn the cheese. Now you can remove the mould, and with boiled but dry lengths of calico, bandage the cheese to keep the sides upright and firm; continue to turn daily. When white patches appear on the outside of the bandage, remove it and leave the cheese to dry in a good current of air. When dry you rub a little lard round the cheese to keep air out and to let it ripen. Use any time after one month, it should keep up to three months, maturing well. Two gallons (9 kg) of milk should produce 1½ lb (675 g) of hard cheese.

Kid Meat

Kid meat, as it has absolutely no fat, is always better marinated for a time in oil and herbs. Rosemary is very nice with it. When cooking, leave some oil with the meat and wrap in foil, cook in an oven for about 20 minutes at 232°C (450°F), not more. If braised in stock and then removed from the bones, it makes delicious pies, especially combined with mushrooms.

Joints from older animals can be pickled in brine and treated like ham, cured in the same way. It is very pleasant boiled and then served cold.

Curing Kid Skins

Kid skins are more often than not thrown away, as it is so costly to have them cured. But it is possible to cure them at home, and many goods can be made from them.

Old method
Mix 1 lb (450 g) of pulverised alum, ¼ lb (113 g) of saltpetre and twice their combined weight of bran. Wash the skin in hot water with strong soap suds. When the skin is damp, spread the bran mixture all over it, 1¼ in (30 mm) deep; fold the skin, fur side out, and leave in a cool place for a week. Scrape the mixture off and work the skin until it is soft.

Second method of curing
The skin should be removed as soon as possible and cleaned. Place the skin in a 25% formalin solution for one week. Remove from solution and wash in clean water until free of formalin. At this stage scrape away any surplus tissue from the skin. Carefully stretch the skin on a suitable sized board, flesh side out, use drawing pins to keep in shape and taut. When dry rub in lancrolin oil to soften, and scrape the skin gently. This step should

be repeated daily until the skin is clean, soft and pliable. The skin is then washed with household detergent to remove surplus lancrolin, and re-stretched to dry fur side out. Lancrolin is obtainable from Watkins and Doncaster, Four Throws, Hawkhurst, Kent. (This curing method was published in the BGS Journal.)

Slippers, handbags and hats, as well as pram rugs made from skins mounted on felt or some thick firm material, are very easy to sell, or make splendid gifts.

A service is now available for tanning skins for rugs, etc., and you will find the address in the appendix. It used to be difficult to tan one skin alone, and home tanning was not always as good as might be hoped.

Cashgora

In the last few years angora goats have been imported into England from New Zealand and Tasmania to be used as wool carriers. As so much cashmere and angora wool is imported yearly, it seems a reasonable idea to try to produce it in England. 'Cashgora' is really making quite an impact in the goat world, with very high prices (thousands of pounds) being paid for imported Angoras. It also seems that because the bony structure of the Angora goat is a little different from that of the milk goat, being shorter from thigh to body, the carcases look more like lamb joints and hence those crosses which are not such good wool carriers are more easily marketable as meat. The prices for the animals have recently become more reasonable as more stock has been bred in this country.

CHAPTER 10

MANAGEMENT

PLAN your day so that you are able to keep to a routine. The time at which you decide to milk will rule the remainder of your daily chores. Some people feed their goats whilst milking, but unless you are milking by machine it is not to be recommended as the dust from the meal, disturbed by the goat nosing the contents of her pail, will get into the milk. The machine, of course, has a closed pipeline and unit bucket.

The idea that food will keep the animal standing quietly does not really work, as she sways gently back and forth in her up-and down search for food. It is much better to feed either before or after milking, preferably before, as the goat will then stand contentedly cudding whilst you milk.

Once having set your times, you should keep to them. A few minutes either way will not matter, but the quickest way to 'dry off' your milker is to do the milking at odd times. In fact, it is the recognised manner of doing so when she is in-kid, but you do not want to lessen the supply and shorten the lactation by being careless on such a basic matter.

POINTS TO OBSERVE

Here are a few points you should observe in your management programme:

- Hay is expensive after so many dry summers, so use it very carefully, but always feed it before letting the goats onto grazing, or give them green feed if stall fed, otherwise they will scour.
- Make absolutely sure that kids' bottles are correctly cleaned. Nothing causes digestive upset more quickly, as bacteria multiply very fast in such a medium as milk residues.
- See that hooves are trimmed at reasonable intervals and do not leave until they turn up at the front like Persian slippers.

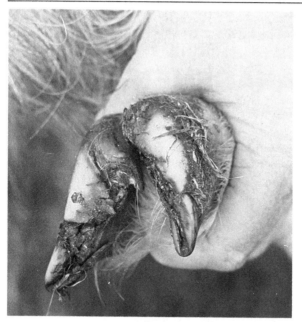

Hoof trimming – before and after.

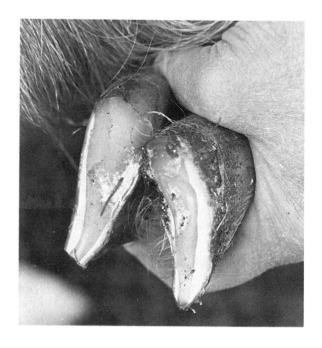

- See that routine tasks such as bi-yearly worming and vaccination against enterotoxaemia are actually done and not put off until a convenient time.
- Make certain that kids are disbudded at four days old, and that male kids intended for meat are castrated so that no accidents happen if by chance brother is left for company.
- If you have the storage space, buy hay and straw in bulk off the field. But it must be stored dry or it will be wasted. On the other hand, order cereal foodstuffs when supplies are running low; they get stale if kept in store too long.
- Remember to register your kids, and to officially transfer any stock sold to their new owners. This must be done on a correct BGS form and a fee for doing so will be sent by the BGS to the new owner.
- Keep your pens dry and clean, not necessarily cleaned out at very short intervals as that wastes straw, but topped up so that the beast lies dry.
- Arrange to get your goats served by the male of your choice well ahead of the time you need to take her. And don't get overstocked—this makes life harassing and causes accidents.
- Inspect fencing frequently; remove nails and snags from walls.

AUTHOR'S ROUTINE

My own routine is:

7.15 am   Feed concentrates to milkers, goatlings and older kids; bottle feed any baby kids; remove all water pails; then feed the males and any other stock.

7.50 am   Refill water pails and place them in the pens, removing empty feed pails.

7.55 am   Prepare strainer with filter pad, place it on top of a churn, putting the churn lid over the strainer centre to prevent dust settling there; the churn is placed in the milking area, and the milking pail is hung on the scale ready to weigh each quantity of milk.

8.00 am   Bring out first milker, who jumps on to the milking stand; udder is washed, she is milked and gets down whilst I weigh the milk and tip it into the strainer over the churn; next animal is let out and first one put back in her pen, and so on until I have finished. Milk churn is removed, strainer taken out and lid put on. The churn

is then placed in the cooler, where running water is set
in being. Pail and strainer are rinsed in cold water, then
filled with hot water and dairy detergent, and left until
I have completed filling the hay racks.

8.55 am All hay racks are filled, lids fastened on and any pens
which require fresh bedding are strewn with straw.
After that I complete the washing up by scrubbing
the utensils with a churn brush, then I rinse them and
fill with sterilant (there are several chemicals, liquid or
powder, recommended by the Ministry of Agriculture).
It is essential, if you want to sell milk or other produce,
to have all articles clean.

10.30 am On fine days goats are let out on to their paddock,
where they stay until 3.30 pm. Young kids, however,
are brought in an hour before they are due for a milk
feed; this prevents scouring, as milk on a full rumen
can cause upsets.

3.30 pm Once the female goats are in, the males are let out. It does
not do to have them out together, as males get annoying
and can harry the females. They are very precocious
creatures. A male kid can give effective service from ten
weeks of age, so at the latter end of the summer
your young female kids could easily be served if left
with their brothers, so play it safe and separate them.

4.30 pm Concentrates are given again, the morning routine
repeated, except that the males are returned to their
house before being fed.

5.50 pm Water pails are filled and the milking routine is repeated.

7.30 pm After hay racks have been filled, kids get final bottles,
and all are left until morning, though I usually have a
quick look at about 10 pm to see that all is well.

In winter the morning routine is the same; but the kids, being
older, do not have morning bottles, just a larger meal ration.

10.30 am Unless it is very wet or frosty, the animals go out for a
run and exercise. This helps to prevent boredom. They
won't want to stay out long, and when let in again can
be given a feed of kale or chopped roots and hay racks
topped up; nothing else until afternoon feed.

4.30 pm Hot soaked beet pulp is squeezed dry and placed on
top of their meal. This makes a good fill-up, as goats,
like trout, feed in the evening and go to bed full. Otherwise routine as before.

Of course this is just my way, and there are many different arrangements.

For those who want to keep goats and are out at work all day, I would suggest no kid-rearing unless it is on their dams, little as I like that method. But it is best to keep the kids away from mother at night. In this way kids have all the daytime milk, and the household the night yield. Some shelter is essential in the paddock or yard, if the animals are to be put out early, before you leave for work. The animals would not be left out in the pouring rain or storm if there was any sudden change in the weather. Such an arrangement would be satisfactory from say April to October. Other months, the goats would be better in their house and let out at weekends or whenever the owner was there to supervise their outings.

CHANGE FOOD GRADUALLY

If any changes are made in the foods offered, always use caution and add a little at a time. Any quick change is a bad thing and can cause digestive upsets. Goats are very conservative creatures and feel very unsure of something new, so give them a little and gradually increase the quantity. Dried beet pulp is usually fed to a milker at the rate of 6 oz (175 g), which when soaked will swell to make about three double-handfuls. Beet pulp is becoming hard to obtain. Pulp nuts are about but you should use less of them. Try a small amount covered with water to see what you need.

The actual times at which you do the various things which have to be done are entirely up to you. But get as long—and as even—an interval between milkings, twelve hours apart if you can. My own milking times are actually 10 and 14 hours apart, not ideal, but it suits me and the milk recorder.

GROW SOME CROPS

I strongly recommend that some land is used for growing crops; you will get much greater returns from your own harvested vegetables. Planting your kitchen vegetables in strips, with some permanent crops and others which are resown each season, can provide a continuous supply of good bulk and add variety to the diet. Most of the vegetables can be used for the household, as well as for your goats, but some are purely animal fodder.

For a piece of land on which you can sow sections approximately 7 ft × 16 ft (2 m × 5 m), starting with the more permanent plants, I would suggest:

2 strips of lucerne
2 strips of herbal hedge-row mixture
1 strip of artichokes
1 strip of comfrey (once established it will continue for several years)
2 strips of thousand head kale
1 strip of marrow-stem kale
1 strip of drumhead cabbage
1 strip of horse tooth maize

Follow the marrow-stem kale by winter oats and vetches, and the maize by stubble turnip mixture and rape and kale mixture, which can be sown in August and stand through the winter.

Lucerne and hedge-row mixture can be sown late summer for cutting next season, when it will be well established. Two or three cuts can be taken each year; and it will stand plenty of manure. Lucerne must be mixed with a fluid which inoculates it, otherwise it does not germinate well, but the seedsman will advise you when you buy your seed.

Herbal mixture is an excellent crop. It contains alsike, agrimony, burnet, sheep's parsley, yarrow, rib-grass, chicory and many other herbs. It is good for 6–7 years and can be cut several times each year, an enormous quantity of bulk for the area it covers.

Rape is quite useful, easy to grow, but it should be used whilst fairly young, not left until almost flowering as at that stage it makes milk taste metallic.

Maize is another greatly liked food, but it must not be sown until frosts are over. It is cut and fed green, not left to form cobs.

Fodder beet and mangolds may also be grown and should be pulled late September and stored in a frost-proof place until after Christmas, when they are ready for feeding. Kales are fed hung, otherwise much is wasted by being dropped on to the ground. Marrow-stem kale is, as its name suggests, fat and juicy stemmed. It can be stripped of its leaves and the stem chopped and fed mixed with the meal; it should be used before Christmas, as strong frost splits and ruins the plant. Thousand-head kale is more hardy and grows on through the winter. If it is left long enough, it will send out masses of side growths in the early spring, hence its name.

Kohlrabi is quite a nice vegetable for house or stock; it grows a bulb like a turnip, but above the ground. The Dutch seed firm Bokker offers a giant kohlrabi seed. The seeds weigh several kilos and do not grow woody. Kohlrabi has a cauliflower-like flavour, is hardy and will withstand frost.

Jerusalem artichokes are most useful tubers; the tops are hairy like comfrey, can be cut on wet days (a shake and all the water drops off) and the plant can be fed right away. The root of the artichoke is a delicious vegetable for the house; cut in halves it is much liked by goats. Just leave one artichoke in the ground for each root you dig, and your next year's crop is ready to grow again. It is a very hardy plant and can be dug as it is wanted. Comfrey can be grown from roots or slips of roots; it gives a very good amount of leaf weight for the area grown, can be cut several times in a summer and the plant increases in size if well fed.

Only by stall-feeding does one realise just how much an animal must eat whilst out grazing. I cannot give any very definite amounts of bulk foods which would be needed as, of course, the size and age of the animal would make its needs vary enormously. However, here are some suggestions as a guide-line:

> Good hay, minimum of 4½ lb (2 kg) daily
> Lucerne (wilted), 5½ lb (2.5 kg)
> Artichoke tops, 4½ lb (2 kg)
> Comfrey, 2½ lb (1 kg)
> Tree leaves, twigs, etc., 9 lb (4 kg)
> Rape, 4½ lb (2 kg)
> Cut grass or herbal mixture, 9 lb (4 kg)
> Kales, 4½lb (2 kg)
> Green maize, 5½ lb (2.5 kg), diced
> Roots, chopped, 2¾ lb (1.5 kg)
> Oats and vetches, 9 lb (4 kg)
> Pea haulm or Kohlrabi, 4½ lb (2 kg)

The goats, having cudded following their early morning feed, should be allowed out into their yard, where they can be given a large rack of whatever type of greenstuff you are feeding. If sunny, they will lie and cud once they have cleared the food. You can then bring them back in after an hour or so, and feed such things as bread waste, pea pods or small weeds (in pails, as these would fall out of racks and be wasted) and then leave them to their hay. Afternoon feeding is as the morning; then hay and another good rack of green food or roots according to the time of year. Finally for the night, hay racks should be filled and topped up with cut greens or cut grass, lucerne or comfrey. Variety is easy to achieve if plenty of these crops are grown. From your own vegetable garden, sprout stalks, cauliflower leaves, blown cabbages, spinach, celery trimmings are all excellent.

COOK POTATOES

Potatoes are not good to feed raw, so if you are able to get stock-feed chats, steam or boil them, and sprinkle them with bran or grass meal; some goats love them, others are not interested. Stock feed carrots are to be had in some places; split them longways, as small round pieces tend to get stuck down the animal's throat. Nettles are very good food and most goats like them, just as they begin seeding. Thistles are also favoured, but not by me if I have to put them into racks!

But, once again, don't give a large amount of any new food at first; go slowly, a little and then more. Enterotoxaemia is the worst ailment a goat can have, and it begins with too much of something. It can happen if the animal manages to get its head into a corn bin, or is allowed out too long on lush pasture without having had enough hay first. It is so fast in its action that quite often the goat can be found dead without having shown any symptoms (see pages 53 and 104).

RATION GRAZING

Goats who did not go out during the winter at all should have their spring grazing rationed, beginning with half an hour, and gradually extending after a couple of days, until they are stopping out for the length of time which you decide is enough.

On first going out they will knock heads and play about for a few minutes, then the herd leader will evidently issue an order, as they will suddenly desist and begin to graze. This goes on for perhaps 45 minutes, they then lie down and cud. All at once, off they go again; this happens three times, then you will find your animals waiting around the gate ready to come in.

Herd leaders emerge, they are not made. I had an elderly goat who was a very good leader, always investigating everything before the lesser fry were allowed to do so. This old lady died suddenly from a ruptured aorta. For two days the goats were at a loss, then with great presence, the old lady's four-month-old kid decided to do the job and simply led them off. Thirty-three goats of all ages, they took her as their rightful leader.

Should you be fortunate enough to have a hillside or common

Various types and breeds on the author's farm, living happily together.

on which your goats can browse, alone or herded, you will be saved a large outlay on fencing and get a great deal of free food.

Paddocks, in which goats are out all day, should be equipped with some sort of shelter against sudden squalls, even if it is only a roof on four posts, or a hut made of straw bales, surrounded by wire netting. This would have to be renewed fairly often, as they will pull the bales apart, just for entertainment.

A water container for drinking water is also necessary. This should not be a pail, which is just something to play with. Drinkers of galvanised iron are heavy enough not to tip easily, and glazed sinks sometimes found in builders' yards are excellent.

POISONOUS PLANTS

The following are poisonous and should be avoided:

Aconite	Box	Fungi
Alder	Bryony	Ground ivy
Arum	Celandine	Gladiolus
Azaleas	Charlock	Gourds
Anemone	Cupresses	Hellebores
Bracken	Daffodil	Hemlock
Black nightshade	Dog's mercury	Holm oak
Broom	Deadly nightshade	Horsetail
Butterbur	Dodder	Iris
Buttercup	Foxglove	Juniper
Buckthorn	Fool's parsley	Knotgrass

Knotweed	Old mans beard	Spindle berry and branch
Kingcup	Poppy	and leaf
Laburnum	Pennycress	Spurge laurel
Lords and ladies	Privet	Thorn apple
Lilac	Potato tops	Tansy
Lupins (yellow)	Pines	Tomato greenstuff
Marsh mallow	Ragwort	Tormentil
Mugwort	Rhododendron	Waterdrop wort
Mulleins	Rhubarb	Yellow flag
Nightshades (all)	Rush	Yew

RESTRAINT

Restraining goats is not at all simple, they are such active beasts. Once you have established a method which suits you and the type of goat you have, they get to know it and train the young ones how to react to their environment.

Cattle or universal fencing is galvanised and has eight strands of wire with oblong mesh, differing in size. The lowest part has small apertures which keep the smaller members of the herd in. I have used this type of fencing for years, with posts 6 ft (1.8 m), driven 18 in (450 mm) into the ground; the gap between the uprights is 10 ft (3 m). The wire is stapled to the posts in at least three places; rolls are 165 ft (50 m) in length, 4 ft (1.2 m) in height.

Chain link fencing is very expensive; it also needs very strong posts as it is extremely heavy, plus straining wires at the corners. It is, however, very effective if high enough.

Never use sheep netting, which is round meshed; this gives convenient toe holds for kids to scramble up.

Cleft chestnut fencing is very dangerous. Small kids can get fast in it and older goats put their heads through the larger gaps and get stuck; they can easily be strangled that way. I would never use it.

High tensile wire, seven strands, will, I am told, keep in older animals, but not the kids.

Electric fencing is very good. Flexinet has seven live strands with poly-twine across the strands lengthways at decreasing intervals and built-in plastic stakes. Not even small kids can get through. It is light, easy to move and can be run from the main electricity supply or a battery. It is all I use now for dividing up larger areas, but I would be loath to trust it for a perimeter, especially close to a road or railway. If used, it needs one wire

3 in (80 mm) above ground level unelectrified, and three strands above it carrying the current. This keeps in all but the most persistent creature. Certainly it is most useful as it is more easily moved than any other type of fencing.

Yards can have concrete block walls, slatted wooden partitions, or some type of welded mesh on strong framing. Wooden hurdles, as used for sheep, would not be high enough; anything less than 5 ft (1.5 m) for females and 6 ft 6 in (2 m) for males is an open invitation to go for a walk.

TETHERING

Tethering is a last resort; of course, in some situations it is the only possible method which can be used, and it is much cheaper.

Running tethers are better to use and less wasteful of space than the round variety; drive two iron pins 20 in (500 mm) into the ground up to their heads, with a taut wire between them; on to that wire thread a short length—3 ft (1 m)—of chain, with a swivel at each end; one goes on to the goat's collar, the other lets her turn round.

This arrangement will allow grazing in strips, and moving twice daily will give a clean area to graze. Goats will not eat grass fouled with droppings.

Tethering can mean quite a lot of work, bringing the animals in when it rains, but to leave a goat tethered out in the rain is cruel. Sheep have a great deal of oil in their coats which repels the wet, but not so the goat, who gets really wet and very miserable. Moreover, it will take all the food she eats just to keep her warm and will leave nothing for production.

CHAPTER 11

THE MALE GOAT

MALES are most affectionate animals, which is unfortunate, as they express their emotion by rubbing the head against their attendant! Since at all times of the year there is a slight odour, which in the rutting season develops into a quite particularly unpleasant and clinging smell, any person must be extremely careful about contact with the male. Always feed him last and have a long voluminous garment to use specially for caring for your stud animal. A mackintosh or plastic coat is suitable as it can be washed down.

Do you really need to keep a male? This is a question which everybody contemplating such a step should ask themselves. If you have less than, say, six milkers or goatlings, or live close to a good male of the right breed, who is not too closely related, I should say, 'No, you don't need to'.

A male must be housed away from his ladies, but not isolated or he will be bored and lonely. The house must be strong, as he will grow into a large powerful animal. He must be well away from the dairy area, otherwise your dairy produce might get contaminated by the penetrating odour.

STARTING WITH A KID

Having made up your mind that a male is necessary, procure the best you can on pedigree, having looked at as many of his female relatives as possible. To begin with a kid is simplest, which means you train him to your methods and he is able to share a pen with your female kids for a few weeks before being separated and placed in his own quarters at ten weeks. After that it would be unsafe to leave him in female company. These poor little fellows lead a lonely life, so if you are rearing male kids or neuters for meat, let them go with him. By the time they are large enough to

Ambrosia kids born in 1988 at M Conisbee-Smith's farm,
Gloucestershire.

make useful freezer food, the season will occupy the mind of your
young stud male.

Partitions between males, if more than one is kept, should not
be less than 5 ft (1.5 m) high; the same goes for fencing. A hinged
panel in the outer wall through which water and food pails can
be put into position, is most useful as there is less chance of getting the unpleasant smell on your clothing. Hay racks can also be
made to be filled from outside. As stated in Chapter 4, mangolds,
sugar beet or beet pulp should never be fed to males—the sugar
content is too high. This causes stones to develop and the passage of urine made painful, or at worst impossible, which entails
having the animal put down. High sugar content can also cause
sterility.

Remember that it is useless to keep a male bred from an ordi-
nary household milker; you must have one who will improve on
what you have already. It takes the same time, space and food to
keep the poor one. With a good male your initial outlay will be
greater, but stud fees will soon repay that and your own stock
will be so much better.

Never allow male kids to jump up or butt in play; this may be considered 'cute' but imagine a smelly male placing his muddy hooves on your shoulders on a wet day!

A male kid will begin his working life from six months or even less, so should be kept on his four bottles of milk until four months, then one bottle at night and one in the morning for two more months. Finally, one bottle in the late evening until nine months. This will keep him growing well; even if his appetite for concentrates does get rather unpredictably affected by the sound of bleating goats on heat. The small handful of meal given him as a tiny kid should soon be increased; he grows fast and at six months should be receiving the same ration as a goatling. Increase this still more when he is working—up to not more than 1 lb (450 g) morning and afternoon;—this will be quite enough even for an early born or extra large kid.

Feed the best hay ad lib, as this is his greatest feed, and give him what greenstuff you can manage or let him graze for an hour or more each day if it is fit, once the female stock is indoors.

As a rule, you will find it necessary to change your male every two years; by that time the daughters of his kid season will be returning as goatlings ready for service. It would be a rather risky thing to use him on his daughters. Often it is possible to exchange a buckling with another breeder—another reason for having a good male to begin with.

WHEN READY TO USE

Teach him that he must go into his house when you say so, then he will be used to this and make no fuss. Once he is being used, he will be brought out to serve, then put back. If he is used to it, no fuss will be made, otherwise he may well decide that it is nicer to play around; he then becomes a nuisance to handle. My own males come out, serve, and immediately return. The concrete apron outside the goathouse is used for standing, as in wet weather it is less muddy and slippery and one can see what is happening.

The owner of the served female is then given a Certificate of Service which, providing the person belongs to either the British Goat Society or one of the affiliated goat clubs, means that the kids can be registered in the correct Herd Book. A short time after the first-service the male is brought back and a second service given. This is because the first service may have been ineffective due to the female being nervous after travelling and many goats travel considerable distances to the male. Therefore the second service is more or less an insurance.

Many people, especially those used to dogs and pigs, can hardly believe that service has been given; it is so fast, if you blink you miss it! The male has his feet back on the ground in a very few seconds, but they are very sure in action. Some males like to take their time making a fuss of the lady first, to spin the whole affair out, but the actual thrust is about two seconds in duration.

STUD BOOK ENTRIES

If you are a member of the BGS or one of the clubs, your male can be included in the Stud Book List for a fee. The details required are: name of the male; Herd Book number and initial; date of birth; earmark number; his sire and dam, both with HB numbers; where the male is 'standing'; the fee for his use; whether you are able or willing to transport him, or can board goats sent to him. This list, which is made up each June, is very comprehensive and useful as it goes all over the British Isles.

Kerney Nevada Br Ch GG1796D owned by Mrs L Mudle Small, Golberdon, Cornwall.

CHAPTER 12

SHOWING

THE show is the breeders' shop window and a good yardstick for those less committed. County shows have classes for most breeds of goats. Golden and English Guernseys go into the same class if such a class is provided for, but many shows do not think there are enough of them to require a separate class, in which case they will go into the AOV (Any Other Variety) class which will also include HB, FB, SR and IR goats. This is a bit hard on the small Guernseys as most of the really heavy milkers are AOV, which often confuses newcomers to showing. Angora, Boer, English and pygmy goats usually have separate shows, as they are not classed as milk stock.

Ask the secretary of whatever show you wish to enter for a schedule and an entry form for each animal you wish to show. Most large shows need to have entries in about two months before the show date to get cataloguing, marquees and penning arranged.

If you enter milking goats, you will have to be present the day before the show, as all goats have to be milked by the exhibitors and then, at a specified time, stripped out by the stewards. Thus all start equal.

For most big shows, do not arrive without your caprine arthritis and encephalitis negative certificate, as you will be turned away without it. Stewards check for this on your arrival. The test must be done by a veterinary surgeon two to three months prior to showing and must include all stock being shown.

Give yourself time at the smaller shows to give all your exhibits a good brush, and possibly wash off any soiled spots from white animals.

One-day shows usually give a rosette with the prize money, if any, attached in an envelope, but the larger affairs send a cheque when all results are known; this may take several weeks.

R141 Ambrosia Drachma GG1264D, winner of the Alderkarr Trophy for the highest milk recorded Golden Guernsey lactation ending 1989, owned by M Conisbee-Smith, Gloucestershire.

WHAT TO TAKE

You will need to take pails for food and water, hay racks, kids' bottles and teats, goats' coats, collars, leads, a milking pail for each milker, hay, concentrates, branches and green grass (or something of that sort) and, according to what is in the schedule, straw. For your own use you will require food, bedding, a sleeping bag, some form of heating apparatus for cooking food and an alarm clock for early rising. Don't forget to take the necessary utensils—a saucepan for heating kids' milk, a frying pan for your sausages and a kettle for those never-ending cups of tea and coffee! And see that you have a churn for the milk which the kids will require, matches for your gas ring, and a good torch.

You will, like most goatkeepers, place your sleeping bag on layers of straw in the gangway between pens, to keep an eye on stock during darkness. Some people sleep in their caravans, but this leaves stock unattended all night and much mischief can be done by an escapee.

The day starts early. Milkers are inspected at 6.30 am with full udders; then when all have been looked at, milking begins. All pails are taken to the weighing table, and butterfat samples removed after the milk has been weighed—in the show pails, not yours. Milk is then tipped into your pail and stood aside until every exhibitors' pail has been accounted for. After checking is complete, you will be told when you may collect your milk. Later in the morning, usually at 9.30 am, inspection proper is carried out and placings made.

Smaller kid, male and club shows are usually half-day affairs, and you only need collars, leads and cord to tie the animals up to the stakes or hurdles provided for the purpose. A bundle of hay or some twigs will keep them happy for that time.

GOOD HEALTH ALL-IMPORTANT

But good health is the best preparation; no amount of grooming will get an underfed goat to produce the bloom of a plump sound animal. Hooves must be trimmed well, as standing firmly on all four feet makes a great deal of difference to the stance. Hooves cut badly can make an animal look cow-hocked, and trimmed on the opposite sides make it walk narrow. Really good handling can make a poor animal look good. Get your animal used to leading prior to going out, or you will feel foolish if she jibs at going into the ring. This is forgiven in a young kid, but not in a milker.

MILKING COMPETITIONS

Milking competition results are rarely known before leaving for home, as the milk has to be tested for its butterfat content by the Milk Marketing Board. To win, the animal must have given the largest quantity of milk with adequate butterfat—3% minimum. To get her Milking Star* she must obtain 18 points in the competition, with 3.25% bf in both milkings; for a Q Star, 20 points with 4% bf in both milkings. Protein has recently been included in overall totals. Its effect on overall quantity is complicated.

On arrival, don't be tempted to give your animal racks of lovely fresh grass you collected; watch the old hands, you will see that a rack of clover hay, or other hard hay is given first; this helps the butterfats and without those you simply have no chance.

Do not take the whole thing too seriously. Whether you win or not, it's good to be able to compare your stock with some which are really good. And it's fun too, particularly in retrospect.

Toggenburg goatling, Alderkarr Nygella, owned and bred by the author. This goat has won many first prizes as best kid and then as best goatling at shows throughout Britain and was sold before being shown as a milker. (Photograph by Frank H. Meads)

CHAPTER 13

THE COMMERCIAL HERD

L. F. Jenner

THE past decade has seen many changes in the world of goat farming, and the rise and fall of many a commercial herd. The next decade will no doubt see many more changes, but for a successful enterprise, goat products must be produced in a totally professional manner, as food production in the highly competitive and hygiene conscious world of today can no longer be considered a hobbyist exercise.

Luckily nowadays there is much data and information available to the would-be goat farmer, but every farm, its situation and the needs of every farmer are different, and no amount of reading can compensate for lack of experience. It would not be a bad idea to have practical experience with some form of livestock before committing oneself to the large numbers of a commercial herd. Better still would be a few goats of your own for at least a year or so, as whilst the needs of goats are similar in many ways to that of other farm animals, they also have individual requirements that need to be fully understood.

STOCKMANSHIP

A high degree of stockmanship is necessary to farm goats successfully. Farming is no nine-to-five occupation and a good stockman will think nothing of giving up Christmas Day to tend a needy animal; remember that goats need attention 365 days of the year. Stockmen are born and if you are unable to spot a poorly animal in a large group, all the qualifications in the world will not compensate for this. Most farm animals respond to human contact and kindness, and goats are no exception. It is a scientifically proven fact that happy, contented animals are more productive. A few minutes spent each day observing the stock is never time wasted, it is pleasurable to both man and beast, and any problems can often be spotted early before they get out of hand; it takes years to breed a good animal but often only hours to lose it.

CAPITAL

Goats are no cheap and easy way into farming; quite the reverse, as the costs involved in setting up a dairy unit are considerable, and if an added value product is to be made, i.e. cheese, yoghourt, ice cream etc., this extra expenditure must be budgeted for at the onset. Many fail through lack of capital; the milk tanker does not collect daily and the monthly milk cheque does not arrive on the mat, as with cow dairying. Ideally there should be sufficient funds to sustain losses for at least a year or two whilst markets are being established. Labour must also be calculated for because, with the best will in the world, it is humanly impossible to be a herdsman, product maker, van driver, marketing manager and book-keeper seven days a week, year in and year out.

LOCATION

The location of a goat farm is not an all-important factor. However, the method in which they are going to be farmed will help to determine the suitability of the site. If there is land available it can be an advantage to graze the goats during the summer months. Alternatively one can consider zero-grazing. Goats are very adaptable creatures and, providing their basic needs are catered for, will happily produce milk on any farm or location.

As one-third of the population of England live inside the M25, it would be foolhardy not to consider London as an important market when it comes to selling products, and when choosing a site it may well be worth considering this factor.

MARKETING

Marketing is something which is a full-time job in itself. Think long and hard on this subject. The inability to sell the product is probably the reason for most business failures. It cannot be stressed too strongly that help and advice on this subject will be time and money well spent, for there is no readily available market for goats' products, as most people in the United Kingdom have never tasted goats' milk nor any products made from it. Happily this situation is gradually changing as we become more integrated with the rest of Europe, and providing the quality of the product is high, there is no reason why the goat industry should not flourish in the future. If we in the United Kingdom do not fill this market need, you can be sure that our European partners will.

WHICH PRODUCT? OUTLETS AND DISTRIBUTION

Selling liquid milk in bulk to a processor is without doubt the easiest option, but it is the most vulnerable because if the processor lets you down overnight it is not always easy to find another customer at short notice and it is also the least financially rewarding. Next to be considered could be cartoned milk to supermarkets. This will involve pasteurising and cartoning machines and a refrigerated delivery vehicle for at least twice weekly deliveries. Other options are ice-cream, yoghourt and cheese etc., the latter being my own choice as cheese and wine have always been one of my greatest joys. These last three items all require large capital investment for equipment, along with a great deal of expertise. There are many courses available and much reading material on all subjects is currently on the market. Whichever product you decide upon, it will need proper packaging, advertising material, outer cartons and somebody with the time and ability to sell it; it is no good waiting for customers to come to you, you have to find them. Whether you choose to sell your products through a wholesaler or directly to a retailer, they will need to be delivered in a refrigerated vehicle. Above all, it must be enjoyable for you to make the product and to eat it.

FIBRE

Goats can be kept for fibre production but the breeds that produce fibre – Cashmere and Angora – are not suitable for milking. They produce only enough milk to raise their kids.

Cashmere is the fine 'underwool' produced in small quantities by all goats, but in marketable quantities by carefully selected strains, mostly of Asiatic origin. It is a very fine fibre (the average fibre diameter is 15 micron) that produces high quality, luxury materials. The price obtained for cashmere is high but the yields per animal are low.

Mohair is the fibre produced by Angora goats. This is a lustrous, curly fibre, quite distinct from cashmere, and the yield per animal is considerably higher. Angora goats have been bred to produce only the undercoat so harvesting the fibre is easier than cashmere as it does not have to be separated from the top coat hair. Mohair has an average diameter of 30 micron, kids giving the most valuable product with a micron value in the mid-20s, becoming coarser with age so that the micron value for adults is in the mid-30s. The resulting materials have a high thermal value, are hard wearing, crease resistant, lustrous and can be dyed brilliant shades.

MEAT

Goat meat is eaten widely throughout the world but in many Western countries it has been replaced by lamb. Dairy breeds in the U.K. have been selected for high milk production for many generations with a subsequent deterioration in carcase conformation. Some strains of Angora and Cashmere goats yield carcases of reasonable quality and, quite recently, a South African breed – the Boer goat – has been introduced which is bred for the sole purpose of meat production.

Apart from the Boers, meat production from U.K. goats falls into three categories. Kid meat is a high quality product which is in demand at certain times of the year but production costs, in terms of milk replacers, are high. It is seldom profitable to produce and should only be attempted against firm orders for the finished product. Angora goats of 12 to 24 months of age, by which time the most profitable fibre has been obtained, can be finished to produce a high quality product. It should be possible to develop a market to take up the surplus wethers produced by the U.K. flock of Angoras. The third category are the culls – mostly breeding stock that have come to the end of their productive lives. There is a market to be found for these animals among the ethnic population in this country but the returns are not high.

STOCK

The breed and type of goat that you choose will be governed to some degree by the product you wish to produce. There are four main breeds to choose from: Saanen/British Saanen, Toggenburg, British Toggenburg, British Alpine and Anglo Nubian. The other breeds fall mainly into the realms of the fanciers' animal.

Saanen/British Saanen. These familiar large white goats are placid and docile animals, well suited to the commercial herd. They are also a useful crossing animal. Generally speaking they produce heavier milk yields than other breeds, but their butterfats tend to be slightly lower. Weaknesses to watch for are bad pasterns and 'ragged' udders.

Toggenburg/British Toggenburg. These are a slightly smaller brown and white goat. Generally speaking their conformation is better than that of the Saanen, and their butterfats are slightly higher, thus ideal for the cheese or yoghourt maker. The disadvantage with this breed is that they are highly strung and later to mature, and should not be mated until well over a year old. This is my own choice of breed and I maintain that a good British Toggenburg will yield as much as a Saanen.

British Alpine. This large black and white goat is uncommon in the commercial herd, mainly because of a lack of available breeding stock. They are an excellent candidate for keeping in large numbers and a good honest animal with good butterfats.

Anglo Nubian. This goat can always be distinguished by its roman nose and long drooping ears. Yields are much lower, but much higher butterfats make it a useful animal for the cheesemaker and a good meat breed. It does not generally take kindly to being one of large numbers, and it would never be my choice of animal, although some people do well with them.

At the end of the day, you must like the breed or cross that you intend to spend so much time with. Whichever animal you choose, there are a few very important factors to consider. It is rarely worthwhile buying animals in milk, as they nearly always lose it with the move. A better plan would be to buy weaned kids or in-kid goatlings. Goats are very susceptible to stress and for them to be able to give their best, which they will certainly need to do, as much time as possible must be allowed for them to acclimatise themselves to their new surroundings. It will also give you time to get used to them, bearing in mind that each and every one has an individual character which you will need to become accustomed to. Try to buy animals that have been used to living together, so that a pecking order has already been established. Always try to buy stock from a farmer that has kept accurate and reliable milk yield data; this will allow you to consider each and every animal's worth.

Health and CAE will also need consideration. An honest farmer would never mind you having a veterinary surgeon examine stock before you buy. If possible, buy stock from a Ministry CAE accredited herd as the disease can dramatically reduce milk productivity. Remember that if you are starting from scratch you might as well begin with good sound stock. Never begrudge paying the extra for the best, as in the long run it is the cheapest; it costs almost as much to feed a bad animal as a good one. Once a herd has been established for a year or two, there can be no excuses for animals that do not give 1,000 litres of milk in 305 days. Under no circumstances buy animals with horns as they will be more trouble than they are worth.

HOUSING

Goats do not need palatial surroundings, but what they do require is a dry, draught-proof, well ventilated, bedded area. As a

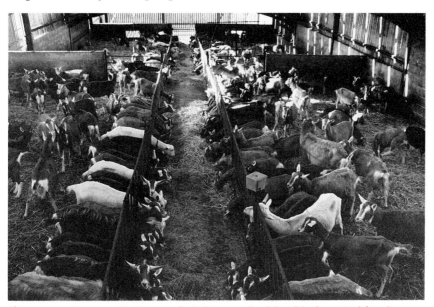

Forage and concentrates are fed down a double-sided central feed passage, allowing plenty of head space at the barrier.

rough guide, groups of animals need a minimum of 17 square feet (approximately 1.6 square metres) per animal and sufficient head space in which to feed. Allow 20% more head space at the feed barrier to alleviate bullying whilst feeding. I am a firm believer in a central feed passage, as hay racks are notoriously wasteful. Make sure all bedded areas can be mucked out by tractor. Try to site water troughs outside the bedded area, with access through a hole in the barrier to prevent the goats from fouling the water.

FENCING

If the goats are to be allowed to graze, weldmesh sheep fencing is adequate, with a high tensile steel wire as a top line. Posts should be no more than two and a half metres apart. A single strand of electric wire along the top of the posts will be a deterrent to climbers, and there will always be some.

BREEDING

Of one thing you can be certain: providing you have taken care with your choice of male and your base stock is of sufficient quality, the animals that you breed yourself should prove to be the best. If you are incapable of achieving this, the success of your whole enterprise will be in doubt. The commercial goat is kidded annually, the goal being 305 days in milk and 60 days dry. With the use of sponges or light treatment, all-year kidding can be achieved to give continuity of milk and consequent reliability of sales. Whether you choose to use artificial insemination or natural mating, remember the male is half your herd, and only keep replacements from females that have already proven themselves. Even then be ruthless and prepared if necessary to cull 50% of the herd.

ARTIFICIAL INSEMINATION

Artificial insemination is merely another method of introducing live viable semen into the female reproductive tract; from there on, a normal pregnancy occurs. There are no myths or mysteries and it is quick, efficient and painless.

Semen can as far as we know be stored in liquid nitrogen almost indefinitely, and from each ejaculate vast numbers of females can be inseminated. We cannot afford to ignore the dramatic improvements that have been achieved in the dairy cow industry in the last thirty years. Milk yields have doubled solely as a result of being able to store proven genetic material. Another bonus of AI is disease control as there is no physical contact between male and female.

It is my strong belief that in the fullness of time AI will be cheaper, as indeed it is in many cases already, and will be widely used not only in this country but throughout the world.

REARING

Kids should be left with their dams for four days after birth. This will ensure that they have had sufficient colostrum and also encourage the dam's milk. Leaving kids on the dam for too long will cause problems later, when training them to drink for themselves from a lamb bar, etc. Never be tempted to rear more kids than you want for herd replacements yourself, unless you have definite orders for them for either meat or breeding stock. There is presently a very limited meat market, and if you are not careful you will end up selling them for less than they cost to rear. Rearing unnecessary stock is a very large drain on resources and should be avoided wherever possible.

As soon as kids are two-thirds of their target adult body weight they can be mated, so they should be reared with this in mind. The earlier you can get them into the parlour the better. They should be fed a good milk replacer for eight weeks which, providing they are eating sufficient solids, should be adequate. Milk replacer is expensive. They should also be given good quality ad lib hay, and encouraged to eat as much bulk as possible.

FEEDING

As stated earlier I am a great believer in a central feed passage with a barrier either side. This needs to be of concrete for easy cleaning and a minimum width of five foot six inches, which is wide enough if necessary to feed big bales. This will also enable you to inspect the stock at a glance. Kids and males should always be fed best quality ad lib hay, but the milking stock can be fed a combination of half hay and half spring barley straw, providing their dairy ration has a protein value of not less than 18% and, just as important, is the metabolisable energy (ME) value of not less than 13. Uneaten straw can be thrown in as bedding. The milkers are fed in the parlour twice daily according to yield. They should also be fed at lunch time and last thing at night (whilst doing the late night check) down the central feed passage, the principle being to space out the feeds to prevent scouring. At about ten days of age kids can be introduced to dairy cake, which must not be more than seven millimetres in diameter, and remain on this diet every day of their lives, albeit receiving only a token amount when they are dry. This ensures that the rumen maintains an even bacterial balance and eliminates unnecessary stress on the animal. The lunch time and evening feeds do not need to be expensive dairy cake; grass or lucerne cubes or a sugarbeet-based coarse mix is adequate.

MILKING

The commercial milking parlour should always be electric, either a bucket system if numbers are not too large or, better still, receiver jars, thereby making recording easier, with a direct pipe-line linked to a bulk tank via a filtration unit, which should be maintained at no more than 5°C. The walls should be rendered and painted to facilitate cleaning which is a legal requirement. There must be sufficient hot water to circulate around the whole system, using a recognised equipment cleaner. You will need some method of restraining the goats whilst being milked; ideally this should be a cascading yolk system.

I prefer to milk the goats from the back. Always teat dip after milking as this will help prevent mastitis. Another useful hint is to keep the animals up after milking for about 15 to 20 minutes. This can be achieved either with fresh bedding which they will always pick over or with some fresh hay or barley straw at the feed bar-rier. Mastitis screens and one-way valves are well worth fitting to the units for the small outlay involved. The goats are milked twice daily at as near to 12 hour intervals as practicable. Milking should be a quiet, stress-free, pleasurable time for both operator and animals alike. In my opinion it is the most rewarding job on the farm.

An electrically operated 12:24 milking parlour, with emphasis on hygiene, efficiency and comfort for both the operator and animals.

One-way valves can help to prevent the whole udder becoming contaminated if only one half is infected.

Mastitis screens enable the operator to spot early signs of mastitis at a glance.

HYGIENE

The hygienic production of milk and milk-based products is something on which there can be no compromise. Remember that one unguarded moment causing a breach in your hygiene standards could cost you your business which you have worked so hard to build. Both your milk and your products should be regularly tested by an independent laboratory for Total Bacteria Count (TBC), coliforms and, in the case of cheese, listeria. Supermarkets will automatically insist on this. It is your legal obligation to produce a wholesome product, and the majority of dairy products in this country are pasteurised. It can only add credibility to goat products if they are also treated this way. The hygienic production of milk starts with a healthy animal and attention to detail must be observed in every stage thereafter.

HEALTH CARE

Firstly enlist the support of an interested veterinary surgeon. Do not wait for an emergency, but establish a relationship that can be relied upon to provide instant advice and assistance. Herd health is dependent upon a regular and routine worming and vaccination programme, and with experience you should be able to assess the general health status and condition of the entire herd at a glance. Feet will need regular trimming; this can be made easier with the help of a crush. Probably the biggest cause of death and loss of milk production is scouring; and this should be constantly watched for and treated almost before it happens. This is when your stockmanship and trained eye will prove invaluable. Although there are many diseases that cause scouring, animals most at risk will be those under most stress, i.e. at the peak of lactation or at or soon after kidding.

SUMMARY

The British dairy goat has quite rightly earned the reputation of being the best in the world. This has been largely due to generations of breeders, the majority of whom have been members of the British Goat Society. Their aims have been to breed a quality milking animal with good conformation and true breed characteristics, backed by meticulously kept records, and for this we owe them a vote of thanks. It is from these animals that the commercial farmer is now able to draw his stock.

No longer is the goat a poor man's cow or a diversification animal, for she has earned her rightful place to be recognised alongside other farm livestock. She is the most proficient milk producer, giving her own weight of milk in ten days at the top of her lactation, compared with a cow who will give little over half that amount. This she can achieve on rougher grazing as she is a better converter of food.

With a modern professional approach and attention to detail, products from this endearing creature will, I am sure, become commonplace in this country, thus ensuring a future for the goat and a profit for those who are fortunate enough to work with them.

KEEPING YOUR GOAT HEALTHY

John G. Matthews

THE British goat is essentially a very healthy animal. We are very lucky that many serious diseases which affect goats in other parts of the world, including many EC countries, are absent from the UK. This means that most diseases which occur in goats are a result of poor management such as overcrowding, poor hygiene, overfeeding, unsuitable feeds, poor foot care, poor milking technique or stressful situations. Good stockmanship will not only produce more contented animals, giving better milk yields, but will also avoid veterinary fees. Routine procedures such as vaccination and worming will not in themselves keep goats healthy and are not a substitute for good husbandry, but should be used as part of an overall management system by goatkeepers to keep disease at bay.

ROUTINE VACCINATION

All goats should be vaccinated to protect them from the various diseases caused by *clostridial bacteria*, in particular enterotoxaemia and tetanus.

Enterotoxaemia (Pulpy Kidney Disease)

This is an acute, often fatal, disease affecting all ages of goat, and often those which are thriving best. It is caused by epsilon toxin produced by the bacterium *Clostridium perfringens type D*, which is present in the intestine of most normal goats and given favourable conditions grows rapidly.

The main danger period is the first few days after any change of pasture or diet, e.g. a change from poor pasture to rich pasture or a higher level of concentrates. It is most important that all changes of diet are made gradually.

104

Many affected goats are found dead in their pens or in a terminally shocked condition with convulsions. Less acutely affected goats show severe abdominal pain (see colic, page 136), shock and diarrhoea often with blood and mucus.

Prevention is much easier than cure. Seek immediate veterinary help for any animal with depression, diarrhoea and colic pains. But many animals will die regardless of treatment.

Tetanus

This is caused by a neurotoxin produced by the bacterium *Clostridium tetani*. Bacterial spores enter through wounds following disbudding, castration, kidding, ear tagging, shearing etc., resulting in signs of the disease 4–21 days later.

The toxin affects the central nervous system producing erect ears, an elevated tail, extended neck and a typical rocking-horse stance. The affected animal has difficulty in opening its mouth and thus eating and drinking. Eventually it becomes recumbent and dies.

Seek veterinary help immediately if tetanus is suspected. Some animals can be saved by a combination of intensive medical care and supportive treatment with forced feeding and fluid therapy.

Other Clostridial Diseases

Clostridium perfringens types B and C can cause an acute, often fatal, haemorrhagic diarrhoea in kids under 3 weeks of age. Other clostridial diseases which affect and kill goats but which are extremely rare in the UK include braxy, black disease and clostridial wound infections.

Vaccination Programme

Only a limited number of vaccines like Lambivac (Hoechst) are licensed for use on goats. Many goatkeepers, however, use sheep vaccines such as Heptovac (Hoechst). If these are administered to milking goats, milk should not be used for human consumption for 7 days.

Vaccination appears to provide poorer protection in goats than in sheep, so more regular vaccination is required, and it is generally better to use a 4-in-1 vaccine, like Lambivac which protects against *Clostridium perfringens* types B, C and D and tetanus, than 5-in-1 or 7-in-1 vaccines, which, although covering more diseases, may produce lower immunity. However, in some areas, particularly where liver fluke are prevalent and other rarer clostridial diseases are known to exist, a 7-in-1 vaccine may be necessary. Your veterinary surgeon can advise you.

Primary course—2 doses, 4–6 weeks apart.
Boosters—every 6 months.
Kids from vaccinated does—start at 10–12 weeks of age.
Kids from unvaccinated does—start at 2–4 weeks of age.
Kids which have not received goat colostrum—start at 2–4 weeks
  of age.
Pregnant does—booster 2–4 weeks before kidding.
Dose—2 ml.

Administration
Thoroughly shake the vaccine. Give by subcutaneous injection (under the skin), either 2–3 inches behind the ear on the side of the neck or over the chest wall behind the shoulder.

Lumps at the injection site
These often occur with goats. Occasionally they are abscesses due to dirty injection techniques, but usually they are sterile reactions to the vaccine which may take 6–12 months to disappear. Some goats are particularly prone to these reactions and certain vaccines are more prone to produce them.

ROUTINE WORM CONTROL

Gastrointestinal parasites (roundworms) are a common cause of disease and economic loss in goat herds (see page 133). It is important to seek veterinary advice to formulate a control programme for your own herd, as no single control system will be suitable for all goat systems. Goatkeepers waste large amounts of money each year by worming at the incorrect time; using unsuitable products; under dosing caused by inaccurate estimation of the goats' weight; loss of milk production in goats carrying heavy worm burdens.

Regular egg counts on faeces samples taken just before each worming will give an indication of the level of infection acquired since the previous treatment. Egg counts should be carried out once or twice yearly.

Remember that newly kidded goats, kids and debilitated animals are the most susceptible to worms, and control is therefore most important for these groups. Good husbandry and a high plane of nutrition reduce the effect of parasites.

Control Systems

All control systems rely on a combination of:

- *Maintaining clean pasture.* If possible, rotate paddocks from year to year. Grow crops such as kale and re-seed, so clean pasture is available each spring, or leave pasture ungrazed until mid July so that overwintered larvae die off. Keep clean pastures for kids.
- *Using anthelmintics (wormers) correctly.* In the typical smallholding situation, clean grazing will be very limited or nonexistent, so anthelmintics are essential to control worm build-up. Because goats develop only poor immunity to worms, adult goats as well as kids need regular anthelmintic treatment.

Control of stomach and intestinal worms

Animals grazing clean pasture in the spring
- Worm in the spring at kidding time.
- Worm again mid-July and move to clean pasture, such as silage or hay aftermath or pasture not grazed in the spring.

Animals grazing contaminated pasture in the spring
If clean pasture is available later in the year:

- Worm in the spring at kidding time.
- Worm again mid-July and move to clean pasture.

If no clean pasture is available in the year:

- Worm in the spring at kidding time.
- Worm every 4–6 weeks from spring to autumn.

Zero-grazed herds
It has been shown that worms can develop in deep litter systems, so even herds which are not grazed outside should be wormed at kidding time and also again at mid-lactation in some cases.

Which wormer to use
Wormers or anthelmintics fall into various different drug groups. To avoid resistance developing, it is best to change *each year* to a drug *in a different group.* Changing within a season is likely to promote the emergence of multiple resistance in parasites. There are basically only three groups of broad-spectrum goat wormers (names of active ingredients are shown on the pack):

- *White drenches*—albendazole, febantel, fenbendazole, mebendazole, oxfendazole, oxibendazole, parbendazole, thiabendazole, thiophanate, e.g. Valbazan (Pfizer).
- *Levamisoles*, e.g. Nilverm (Mallinkrodt).
- *Ivermectin*, e.g. Oramec (MSD).

The choice of drug will depend on:

- Cost, which includes size of container (your veterinary surgeon will probably give you sufficient for one or two goats).
- Efficiency; whether it is active against inhibited larvae or tapeworm.
- Ease of administration, in the form of a drench, paste, powder, pellets, or an injection.
- Milk withholding times, which vary from 0–28 days.
- Meat withholding times, which can be up to 28 days.

It is important to use drugs which are specifically licensed for goats wherever possible. The milk and meat witholding times, i.e. the times before which they can be used for human consumption, will then be clearly stated. If nonlicensed products are used there are statutory withholding times of 7 days for milk and 28 days for meat. Unfortunately, the limited number of licensed wormers available means that if wormers are to be used in rotation, it will be essential to use nonlicensed products. The timing of worming is then critical to avoid excessive wastage of milk.

Remember that it is essential to dose animals accurately. If possible weigh animals before treating. Weigh bands provide a reasonable estimate of weight. Doses for goats are not always the same as for sheep; goats often need higher doses. Consult your veterinary surgeon or the drug manufacturer if in doubt.

Control of liver fluke

In certain areas of the UK (generally the wetter, western areas), it is necessary to treat for liver fluke. Other goatkeepers and your veterinary surgeon can advise if you are in a problem area.

The liver fluke (*Fasciola hepatica*) is a parasite of the liver of sheep, goats and cattle. Chronic fluke disease is commonest in goats as a result of liver damage caused by migrating fluke and blood loss caused by adult flukes. Affected animals lose condition, become anaemic and progressively weaker and may die.

The mud snail (*Limnaea truncatula*) acts as an intermediate host for completion of the life cycle of the fluke, so the disease occurs only in wet, poorly drained areas where the habitat is suitable.

If clean pasture is available, graze fluke-affected pasture before July and move to clean pasture for the rest of the grazing season.

If clean pasture is not available, in mild or moderate fluke years treat with a flukicide in October and January or May, depending on your grazing system; in severe fluke years add 2 further treatments 6 weeks after the October treatment and 6 weeks after the May treatment.

Seek veterinary advice on the correct timing of treatment and the products to use, as most flukicides are not specifically licensed for use in goats.

Control of tapeworms

A number of tapeworms commonly inhabit the small intestine of the goat, including the Moniezia species. It is possible that large numbers could cause a blockage of the intestine in kids, but they are generally believed to cause no problem in older goats. Goatkeepers are only likely to be aware of the presence of tapeworms if a large adult worm is passed.

Most wormers are not effective against tapeworms so veterinary advice is necessary if treatment is required.

The larval stage (*Coenurus cerebralis*) of the dog tapeworm, *Taenia multiceps*, can develop in the brain of the goat resulting in nervous signs known as 'gid'. Similarly cysts of the carnivore tapeworms *Taenia hydatigena* and *Echinococcus granulosus* cause liver damage. Dogs become infected by eating uncooked carcases containing tapeworm cysts and pass eggs in their faeces. Grazing goats become infected by eating contaminated herbage. All dogs that come into contact with goats and sheep should be regularly wormed. Sheep and goat offal should be cooked before feeding to dogs.

ROUTINE TREATMENT FOR EXTERNAL PARASITES

Lice

All goats should be routinely treated for lice. The main species found on goats are *Damalinia caprae*, the biting louse, and *Lignognathus stenopsis*, the sucking louse. These can cause irritation, scratching and damage to the hair, and in severe infections they can reduce milk production or depress weight gain and cause anaemia.

Lice are spread from animal to animal with the entire life cycle being spent on the goat, so there is no need to treat the goathouse. Eggs are laid and attached to the hairs. The eggs hatch and proceed through three nymphal stages before the adult louse appears; this takes between 2–4 weeks. Adult lice are visible to the naked eye in the coat.

Treatment: Louse powders are cheap, but are no longer licensed for use on animals in the UK. Pour-on preparations containing the synthetic pyrethroid cypermethrin such as Parasol (Ciba) are more expensive, but easy to apply and can be safely used for milking goats provided they are treated immediately after milking. The preparations give up to 16 weeks of lice control. Regular use should eradicate biting lice from a herd. Pour-on louse preparations will also treat ticks and blowfly strike, but will not eradicate sucking lice. Other drugs such as Taktic (Hoechst) will kill sucking lice, but are not licensed for goats, so 7 day milk withholding times apply.

Ticks

Ticks are common external parasites found in certain areas. They feed periodically on the blood of sheep and goats, then drop off and spend most of the year in thick undergrowth. They do not cause skin disease but can spread several diseases when they bite the goat.

Blowfly Strike (Maggot Infestation)

This is much less common in goats than sheep because goats have a much thinner coat, but open wounds or badly soiled coats in the summer may attract flies which lay eggs. These subsequently produce maggots who feed on the surrounding tissue causing intense irritation, discomfort and loss of condition. Local treatment should be carried out with topical powders such as Negasunt (Bayer) or a pour-on preparation like Parasol (Ciba) and as many maggots removed as possible. Treatment with a pour-on preparation such as vetrazin (Ciba) will help prevent a recurrence. (Bayer; not licensed for use in milking goats.)

ROUTINE TESTING FOR THE CAPRINE ARTHRITIS ENCEPHALITIS (CAE) VIRUS

One serious disease which is present in this country is CAE. However, prompt and sensible control measures adopted by most goat breeders in the late 1980s stopped the rapid spread of the disease and eliminated many infected animals. The majority of registered goat herds are now regularly tested for CAE and are clear of the disease. *Only buy stock from herds tested and pronounced negative for CAE.*

Tests are performed on blood samples taken by a veterinary surgeon and sent to the laboratory to check for the presence of antibodies. Many goatkeepers have the tests done annually without membership of any formal scheme. Others are members of one of the following:

- *The British Goat Society's Monitored Herd Scheme* provides goatkeepers with a standardised regime for regularly testing their herds. Records of the movement of goats on and off the premises are kept, but no restriction on movement is made. Details of the scheme can be obtained from the British Goat Society.
- *The Sheep and Goat Health Scheme* also requires regular testing of goats but, in addition, places restrictions on the movement of goats: goats can be mixed only with goats from other herds in the scheme. This raises difficulties at present for most dairy goatkeepers as it limits the movement of animals for mating and showing. Details can be obtained from Sheep and Goat Health Schemes, PO Box 604, Milton Keynes, MK6 1ZZ.

All goats should be tested annually.

In parts of the world where CAE is well established, such as France and the USA, the disease occurs in four clinical forms:

- *Arthritis.* This is usually seen as firm swellings of both carpal (knee) joints, although any other joint may be affected.
- *Pneumonia.*
- *'Hard Udder'.* Recently kidded goats affected with CAE develop a firm udder with little milk production and a reluctance to let milk down.
- *Encephalitis.* This is generally seen in kids less than 6 months of age, although occasionally in adult goats causing paralysis.

Many goats are infected with the CAE virus although they do not appear clinically ill, as only a percentage of infected goats will develop the disease. Thus, apparently healthy goats can carry the virus and infect their own offspring and other goats. Once a goat is infected with CAE it remains infected for life, although levels of antibody can fluctuate so that detectable levels of antibody are not always present. **One negative test does not guarantee freedom from the disease.**

The disease can be transmitted by close contact between goats and by the passage of body fluids, e.g. saliva, but the main means of transmission is through colostrum or milk, particularly through the feeding of pooled milk to kids.

Pooled milk should never, under any circumstances, be fed to kids because one infected goat could then infect a whole generation of kids.

FOOT CARE

The majority of cases of lameness in goats involve the foot and most foot lameness results from poor foot care. Conformational problems such as weak pasterns or cowhocks will predispose the goat to uneven hoof growth, and environmental factors such as excessively wet conditions, which soften the horn, may lead to excessive horn growth and increased susceptibility to infection.

There are three principles involved in the care of the foot:

- *Keep the feet dry.* The goat originally lived on hard, dry surfaces like rock and scree. Exposure to moisture for long periods softens the hoof and increases the risk of conditions such as footrot (see page 148). Whenever possible, goats should be housed during prolonged wet periods or kept on concrete rather than grass. Wet bedding is equally bad for the feet. Housed goats should be kept on deep, dry, straw bedding.
- *Keep the feet moving* on hard surfaces such as concrete.
- *Keep the feet properly trimmed.* Because goats are generally kept in conditions which do not keep the feet naturally worn down, regular foot trimming at about six weekly intervals is essential to remove surplus wall horn (see page 74).

THE SICK GOAT

Even in the best run herds, animals will occasionally fall ill. Most goatkeepers will soon learn to spot unwell animals by slight changes in behaviour, general signs of ill health such as loss of appetite, lethargy, a staring coat, puffed-up face, cold ears, drop in milk yield, failure to chew the cud or more specific signs such as lameness, coughing or diarrhoea.

Whenever a goat is off-colour, check its temperature, pulse, respiration and rumen movements.

The normal **temperature** of the goat is about 102.5–103°F (39–39.5°C). Shake the thermometer down, lubricate it with obstetric lubricant, insert it carefully into the rectum and leave for 30–60 seconds before reading. **Respirations** can be counted by watching the goat's flank; the normal rate is 15–20 per minute. The **heart rate** can be felt by placing your fingertips on both sides of the lower rib cage. Count the number of heartbeats for 1 minute; the normal heart rate is 70–95 per minute. The **pulse** can be taken by feeling for the big artery on the inside of the upper rear leg. **Rumen movements** can be felt by placing your fingertips, palm or fist on the left flank between the ribcage and hind leg. There should normally be 1–1.5 movements per minute. Practise these measurements on a normal goat, so that when faced with an emergency you know what to do. It is of much more use to your veterinary surgeon if you can report that 'Daisy' is quiet, not cudding, has a temperature of 104°F, a slightly increased respiration rate and a cough rather than just saying that 'Daisy' is not very well today.

It is always difficult for a novice stock-keeper to know when to call the veterinary surgeon. There is always a conflict between wanting to do the best for the animal yet not wasting the veterinary surgeon's time (and incurring a large bill!). In the following sections I have tried to indicate at what stage the veterinary surgeon should be called, but it all boils down to the experience and confidence of the individual goatkeeper. Most veterinary surgeons would rather have an early telephone call to discuss a possible problem than be called out at two in the morning!

Although goats are generally healthy animals, they do not make particularly good patients and good nursing is essential for a quick recovery. Most goats enjoy human company, and constant attention to a very sick goat may mean the difference between life and death. The sick goat should be kept warm in a well-strawed, draught-free pen, preferably on its own but within sight and sound of other goats, and rugged up (use a sack or blanket if a proper goat rug is not available). The appetite can be tempted with a variety of choice titbits. There is usually a better response to twigs and leaves (try ivy in the winter), warm sugar-beet or molassed water than to bran mashes and oatmeal gruels.

Try to ensure adequate water intake; warm water or water flavoured with cider vinegar, orange juice or molasses will be preferred to cold water. If necessary dose every 4 hours with glucose water made up with a tablespoonful (15 g) of glucose to half a pint (284 ml) of water.

Goats are very susceptible to **pain**, and aspirin is a very good painkiller for them. It is best given in soluble form, dissolved in a small amount of water, administered with a syringe. The dose for adult goats is 6 tablets, 3 times daily and for kids 1–2 tablets 3 times daily.

THE VETERINARY FIRST AID BOX

Antiseptic/disinfectant
Antiseptic wound powder
Antibiotic ('purple') spray
Surgical spirit
Udder cream
Udder wash
Obstetric lubricant
Bloat drench
Epsom salts 500 g
Vegetable oil
Kaolin mixture (scour mixture)

Bottle of calcium borogluconate
    with magnesium, phosphorus
    & dextrose
Soluble aspirin
Clinical thermometer
Scissors, curved 6 inches
Bandages 5 cm × 5 cm
Adhesive plaster
Kidding ropes
Lamb stomach tube and syringe
Large syringe (50 ml)
Torch

BREEDING PROBLEMS

Breeding problems normally fall into one of four categories:

- Failure to come into season (anoestrus)
- Irregular seasons (irregular oestrus cycles)
- Regular seasons (regular oestrus cycles) despite the doe being served by the male
- Difficulty at service.

Failure to Come Into Season (Anoestrus)

This may be due to:

- *The time of year.* Most goats will only show signs of heat between September and February (see pages 39 and 120).

- *Pregnancy!*
- *Poor heat detection*, (see page 39).
- *Poor nutrition*. Both overfat and underweight goats may fail to come into season. Deficiencies of minerals such as cobalt, selenium, zinc, iodine and copper and vitamins B_{12} and D may also cause infertility.
- *Stress*. Goats which are upset by bullying, a change of home or painful conditions such as lameness may fail to come into season.
- *Pseudopregnancy ('cloudburst')*, (see page 117).
- *Intersex*. The mating of two polled goats will result in a percentage of intersex, sterile animals. Female intersexes are genetically female but externally can range from an apparently normal female to male in appearance. Some animals are obviously abnormal at birth with an enlarged clitoris but others may reach maturity before being detected.
- *Freemartin*. Unlike in cattle, freemartins, i.e, sterile females which have been born as a twin to males and have received male hormones, are uncommon in goats but do occasionally occur.
- *Ovarian malfunction*. Ovarian inactivity, with the ovaries failing to produce follicles or with a retained corpus luteum ('yellow body'), will result in anoestrus.

Treatment: Correct any obvious problem such as overgrown feet or bullying; check the diet to make sure it is adequate in protein and energy; supply a mineral lick together with specific mineral supplements if in a deficient area. Exposure to a male goat or a 'billy rag' (a rag rubbed on a male's head and sealed in a jar) may stimulate some females to cycle.

If there is the slightest possibility of pregnancy always use an efficient method of pregnancy diagnosis (see page 118) before the administration of any drugs which might cause abortion.

Seek veterinary help as early in the breeding season as possible. The veterinary surgeon can use a number of diagnostic methods—laparascopy, hormonal assays, ultrasound—to diagnose the cause of anoestrus and there are a number of drugs which help to correct the problem such as prostaglandins and follicle stimulating hormones, but these will only work *within* the breeding season.

Irregular Oestrus Cycles

Normal, non-pregnant, mature females cycle about every 21 days during the breeding season. Abnormalities may result in short or long oestrus cycles.

Long oestrus cycles
These may be due to:

- *Early embryonic death.* Early death of the embryo in a goat which was mated and became pregnant will result in a return to oestrus following the reabsorption of the embryonic material. A percentage of these animals will *not* return to oestrus but will develop a false pregnancy (see page 117).
- *Silent oestrus.* Some goats cycle early in the season and then show no signs of heat for some months although they are still cycling 'silently'.

The treatment is to serve the goat when she comes into season.

Seek veterinary help if the irregular cycling recurs or if the goat fails to come into season by 42 days after the last season was observed.

Short oestrus cycles (less than 21 days)
Short cycles of about 7 days are common at the start of the breeding season. The first season (known as a 'false heat') often does not result in ovulation but the second 7-day heat is usually fertile and pregnancy will result after service at this time. Short, infertile cycles occasionally also occur at the end of the breeding season. Kids commonly show short cycles during their first breeding season, but at other times short seasons are abnormal. They may be caused by:

- *Stress* due to, e.g. bullying or recent mixing in large groups, which often causes short cycles of around 7 days.
- *Ovarian cysts* may result in cycles as short as 3 days or even continuous heat. The goat may show male-like mounting behaviour.
- *Inflammation of the reproductive tract,* e.g. vaginitis.

Seek veterinary help immediately if there is no obvious reason, e.g. stress, for the short cycles. Treatment may involve further investigation, e.g. laparoscopy or surgery and the use of hormones which will only be successful during the breeding season.

Regular Oestrus Cycles (failure to become pregnant)

This may be due to:

- An *infertile male.*
- *Stress*, e.g. transporting the female to the male.
- *Service at the wrong time (delayed ovulation).* Ovulation normally occurs about 36–48 hours after the start of oestrus. Although there is little scientific evidence that delayed ovulation causes fertilisation failure, in practice a 'holding injection', of Receptal (Hoechst) or Chorulon (Intervet) given at the time of service will aid fertility in some animals by stimulating ovulation on the day of service. It may also help if the female is travelling a distance to the male.
- A few goats show signs of oestrus during pregnancy.
- *Inflammation of the reproductive tract*, i.e. metritis or vaginitis.

Treatment: Check that other females served by the male have become pregnant recently; check the female for abnormal vaginal discharges; consider boarding the doe to reduce stress from travelling just before and after service and so that the female can be served every 6 hours whilst in season.

Seek veterinary help if there appears to be a problem of male infertility, the doe returns to service more than twice, or there is an obvious discharge.

Difficulty at Service

Difficulty at service usually arises because the doe is not in season, she is stressed by a long journey, or is scared, there is a persistent hymen or vaginal constriction (uncommon), or a problem with the male such as poor libido (e.g. breeding out of season), difficulty in mounting, thrusting or overwork.

Treatment: Check signs of oestrus (see page 39). If possible avoid travelling long distances to the male and use an experienced male with an experienced owner.

Seek veterinary advice if there is a physical restriction to service or if there is a physical problem with the male.

False Pregnancy

False pregnancies are common in goats which are 'running through' and not mated, and in goats which are mated and conceive but suffer an early (often unnoticed) loss of the embryo.

Instead of dying off, the corpus luteum in the ovary (see page 119) remains active and continues to stimulate the production of the hormone progesterone which fools the goat into thinking she is pregnant; normal cycling stops, fluid accumulates in the uterus so the goat's abdomen becomes distended and her milk yield drops. The accumulated fluid is spontaneously released ('cloudburst') generally sooner than the normal gestation period, although some supposedly 'pregnant' goats carry the uterine fluid beyond full term.

Seek veterinary help whenever the condition is suspected. Once it has been established (see pregnancy diagnosis below), the false pregnancy can be quickly terminated by a prostaglandin injection.

Pregnancy Diagnosis

Accurate pregnancy diagnosis is essential for all herds large and small. Non-return to service is not a reliable method of pregnancy diagnosis because of the seasonal nature of the breeding cycle, the relatively high incidence of false pregnancies and the fact that some goats may not outwardly cycle. Equally unreliable is udder development in first kidders because many non-pregnant young goats are maiden milkers.

The incidence of immaculate conceptions seems to be very high where males and females are kept on the same premises—every year there are cases where the owners swear mating cannot possibly have taken place! Remember, male kids may be fertile from 3 months of age. 'Pregnant' goats that reach full term and turn out to be barren are even more common.

There are several accurate cheap methods of pregnancy diagnosis available:

Oestrone sulphate assay
Oestrone sulphate can be measured in milk or blood from 50 days after service to distinguish between non-pregnancy, pregnancy and false pregnancy. False negatives occasionally occur particularly if the sampling is close to 50 days, so if in doubt repeat the test.

Ultrasonic scanning
Real-time ultrasound scanning is virtually 100% accurate in determining pregnancy and 96–97% accurate in determining twins and

triplets. It is best carried out between 50 and 100 days after service, although it can be used from 28 days.

Doppler ultrasound techniques

The foetal pulse can be detected after about 2 months gestation. Between 60 and 120 days gestation, the method is more than 90% accurate in determining pregnancy but is unreliable in detecting multiple foetuses.

Ballotment

Ballotment of the right flank or ventrally is a time-honoured goat-keeper's technique of pregnancy diagnosis, but it can be extremely inaccurate. Foetal movements can be observed on the right of the abdomen during the last 30 days of gestation but can be confused with rumen movements.

CONTROL OF THE BREEDING CYCLE

Within the Breeding Season

In the goat, the corpus luteum or yellow body is present in the ovary between about days 4 and 17 of the normal cycle and throughout the whole of pregnancy. Corpus luteum-derived progesterone maintains pregnancy. The corpus luteum can be destroyed at any time by an injection of prostaglandin. Therefore prostaglandins can be used to time oestrus, synchronise it, produce abortion and induce parturition. The effect of the injection is generally seen in 24–48 hours, usually about 36 hours.

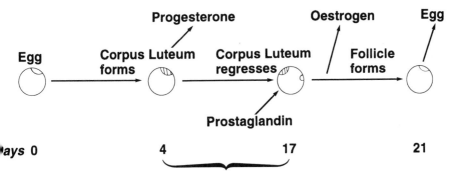

Figure 7   Basic oestrus cycle in the doe.

Progesterone-impregnated sponges such as Chronogest (Intervet) can also be used to control the cycle. These are inserted in the vagina and release progesterone while in place (between 12 and 21 days depending on the type of sponge). During this time the goat is prevented from coming into heat. Heat occurs 12–48 hours after sponge withdrawal.

Extension of the Breeding Season

Progesterone-impregnated sponges can also extend the breeding season. They are most successful when used just before the start of the normal season, e.g. July or August, but can be used in March or April, and with less chance of success even completely out of season between April and July. Your veterinary surgeon can advise on their use.

Decreasing day length is the main factor in initiating the start of the breeding season in autumn. By housing goats in buildings giving 20 hours of artificial light for 60 days from 1st January and then reverting to natural light conditions, goats will be induced to cycle between April and June with peak activity in May. Any males to be used should also be subjected to light treatment.

PROBLEMS CONNECTED WITH KIDDING

Abortion

Abortion is the premature expulsion of the foetus before full term. This can occur as a result of specific infection of the uterus by a number of organisms, e.g. Toxoplasma, Chlamydia (enzootic abortion) and Campylobacter, or as a side effect from some generalised infections, e.g. enterotoxaemia. However, most abortions result from non-infectious causes, usually failure of the corpus luteum (yellow body) either spontaneously or following stress such as being chased by dogs. Abortions can also be caused by drugs, e.g. prostaglandins, plant poisons, malnutrition and where there are multiple foetuses present. Thus, although 'abortion storms' affecting whole herds of goats can occur, generally only individual animals are involved.

Immediate action
Some infectious causes of abortion are zoonoses, i.e. diseases which also affect humans. Gloves should always be worn when handling aborted products. Pregnant women should *never* handle aborted

material and indeed should never assist at any kidding.

Save all the products of abortion, i.e. foetus/foetuses and foetal membranes (placenta), for possible further examination. Any material not required for laboratory examination should be burned or buried and kept away from dogs, cats, foxes or birds. Isolate the aborted doe until a diagnosis is reached and/or uterine discharges have ceased.

Seek veterinary advice after any abortion. If there is no obvious reason for the abortion, arriving at a diagnosis will probably depend on laboratory investigation of aborted material and possibly blood samples from the doe. Some aborting does may be ill but most will not be because abortion occurred either from a non infectious cause or following an infection some time previously.

Vaginal Prolapse

This is the protrusion of the vagina through the vulva in response to excess abdominal pressure in the heavily pregnant animal. The condition appears to be more common in goats carrying multiple foetuses and there may be a greater likelihood in certain families suggesting an hereditary component (due to conformation?). In mild cases the vagina may only protrude slightly when the goat lies down; in more severe cases a large, reddish mass will protrude all the time and there is a danger of the vagina becoming damaged, dirty and infected. The goat may strain continuously because of the increased mass in the pelvis. The condition is likely to recur during subsequent pregnancies.

Seek veterinary help if the goat is straining or if there is danger of damage to the vagina. The vagina can usually be replaced quite easily and the vulva lips sutured to prevent a recurrence. The sutures need removing just before kidding.

Difficult Kidding (Dystocia)

This is generally due to a malpresentation, particularly where a number of kids are present, occasionally because of an oversized single kid and sometimes because the doe is too fat.

Seek assistance from a more experienced goatkeeper or veterinary surgeon whenever kidding seems protracted or the doe has been straining for more than an hour without obvious success. Antibiotics should be administered whenever assistance is given at kidding.

Failure to Kid on Time

The normal gestation period is about 150 days (145–154 days). In general, goats carrying several kids tend to kid early and those with one kid late.

Seek veterinary help for any goat which has not kidded 154 days after she was mated (double check the service date!). In general the longer the pregnancy, the larger the kid and the greater chance of a difficult kidding. The administration of a prostaglandin (see page 119) will induce kidding in about 36 hours (12–48 hours) or a 'cloudburst' in a goat with a false pregnancy.

Mummified Kids

Mummified kids are produced if a kid dies in the womb following an infection or placental failure. It is quite common to have one or two normal kids produced together with a mummified kid. Mummified kids may be retained in the uterus and then expelled a few days after the birth of the normal kids. Occasionally they may be retained for very long periods (several months) and then expelled as bits of bone and flesh. If pieces of kids are left behind it will cause infertility.

Straining After Kidding

Seek veterinary help for any goat which starts straining again hours or even days after kidding has apparently finished. This may just indicate soreness and inflammation of the vagina, but could be caused by the presence of a retained kid. It may also lead to a prolapsed uterus.

Uterine Prolapse

Uterine prolapse, where the entire uterus is pushed out through the vulva, is a rare condition occurring a few hours after kidding and often subsequent to a retained placenta (see page 124).

Seek veterinary help immediately. Before the arrival of the veterinary surgeon keep the goat warm by rugging and the uterus as clean as possible, restraining the goat where necessary and protecting the uterus with a sheet or towel.

Hypocalcaemia ('Milk Fever')

At kidding all goats experience a fall in the levels of blood calcium and phosphorus because of the start of milking. In some the fall in blood calcium is so great that milk fever results. The disease is commonest in young, high yielding first kidders the first few weeks after kidding, but it can occur in late pregnancy, during kidding and at any stage of lactation, particularly in heavy milkers. Older goats are more likely to suffer from hypocalcaemia around and during kidding.

Affected goats show nervous signs ranging from slight tremors and twitching with unsteadiness and hyperexcitability to marked incoordination and/or fits followed by paralysis, recumbency and coma. Some goats may be quiet with a poor appetite and drop in milk yield without obvious nervous signs.

Treatment: Affected goats should be treated with 80–100 ml of calcium borogluconate, with added magnesium and phosphorus, injected subcutaneously.

Seek veterinary help if there is not an immediate improvement or the goat relapses.

Hypomagnesaemia (Grass Staggers)

Hypomagnesaemia is an uncommon condition usually occurring in the winter and early spring in milking goats kept extensively at pasture. It is unlikely to occur in zero-grazed goats with supplementary feeding.

The fall in blood magnesium leads to a rapid onset of the disease; often the animal is found dead. Animals show hyperexcitability, twitching, aimless wandering and later convulsions and death.

Treatment: Inject with 80–100 ml of calcium borogluconate with magnesium and phosphorus subcutaneously, plus 100 ml of magnesium sulphate subcutaneously.

This is a genuine emergency. There will usually not be time for the veterinary surgeon to attend, but veterinary help is essential to check magnesium levels in other animals in the herd and to discuss supplementing their ration usually by feeding calcined magnesite or introducing magnesium bullets into the rumen.

Transit Tetany

Transit tetany is caused by a combined deficiency of calcium and magnesium brought on by stress, particularly by transporting the goat in a vehicle, but also by fear etc. It can occur in dry goats, males and milkers. The signs and treatment are as for hypomagnesaemia.

Vulval Discharge

A light, odourless, reddish discharge (lochia) is normal for about 14 days after kidding. Seek veterinary advice if the discharge is darker and stickier than normal, if it contains pus (indicating metritis) or if the goat is obviously unwell or milking poorly.

Retained Placenta

Foetal membranes should be passed within 12 hours after kidding. Membranes which are still hanging from the vulva after a few hours can often be removed by a **gentle**, even, steady pull. Remember that goats will eat their own and other goat's placentae, so a missing placenta may well be inside a goat's rumen! It is very, very, uncommon to have a retained placenta which is not at least partially visible externally.

Seek veterinary advice if a retained placenta is not passed within 12 hours. After this time contraction of the uterus and closure of the cervix will prevent manual removal.

Loss of Appetite Before Kidding

This may be due to:

Physical restriction on intake
In the latter stages of pregnancy, particularly when several kids are being carried, the enlarging uterus will restrict the capacity of the rumen. In addition, the goat will be less willing to stand for long periods, to wander and browse. Increasing the concentrate ration during the last 2 months of gestation will allow for the increasing nutritional demands of the kids and for the restrictions on intake of bulky food. However, the goat should continue to eat reasonable quantities of both roughage and concentrate feeds right up to parturition.

Pregnancy toxaemia

This occurs when the pregnant goat is unable to eat sufficient food to meet her own nutritional needs and those of her growing kids, resulting in an energy deficit. As 80% of the birth weight of a kid is acquired during the last six weeks of gestation, this is the most critical period, and pregnancy toxaemia usually occurs during the last 3 weeks. It is more likely in goats carrying 2 or more kids, and when there are very large numbers of kids, e.g. 4 or 5, it is often the restricted intake (see page 124) which precipitates the energy deficiency. It is also more likely in overfat goats as large amounts of intra-abdominal fat means less room for the rumen to fill and also because their liver function is likely to be impaired. In other cases, it can be precipitated by starvation or faulty feeding.

The energy deficit results in a low blood glucose level which leads to the doe using body reserves of fat and protein in order to try to correct the problem. The liver converts these reserves into glucose which can be utilised by the foetuses. Ketone bodies are produced as a by-product and accumulate in the blood causing ketosis. Ketone bodies can be detected in milk and urine.

The low blood glucose level depresses the central nervous system and this is exacerbated by the build-up of ketone bodies. At first there is only a slight depression and loss of appetite which worsens the energy deficit; the goat starts picking at food. As the condition progresses, the depression gets worse, the appetite decreases further and nervous signs such as twitching of the ears and muscles of the legs and unsteady movements become apparent. Eventually the goat becomes recumbent, goes into a coma and dies.

Treatment: Encourage the goat to eat and keep her eating. Feed evergreens such as ivy in winter, cabbage leaves, good hay, branches and bark. Give 80–100 ml of propylene glycol (e.g. Ketol, Intervet) by mouth or 200 ml of glucose electrolyte solution (e.g. Liquid Lectade, Pfizer) or 80 ml of glycerol (glycerine) in warm water. Repeat daily. Also encourage exercise.

Seek veterinary help if appetite does not improve in 48 hours or immediately if the goat shows any nervous signs. The patient often does not respond well to treatment and it may be necessary to abort the kids to save the doe's life.

Loss of Appetite After Kidding

Sudden changes of feed should be avoided at kidding. Even a

change of batch of a particular concentrate may result in feed refusal for up to a week. Following a normal kidding, there is generally no loss of appetite and it increases as lactation commences. It follows that any loss of appetite is abnormal and may be due to:

Infection
If the goat is ill from any infective cause, this will decrease appetite. Particular problems after kidding are metritis (see page 124) and mastitis (see page 139). Does are also more prone to digestive upsets and diarrhoea (see page 133). If the appetite remains depressed this may lead secondarily to ketosis (see below).

Laminitis
Laminitis (see page 149) is common after kidding, either on its own or in combination with a toxic condition such as metritis, mastitis or retained afterbirth. The pain in the feet restricts movement and eating.

Ketosis (acetonaemia)
Ketosis is most prevalent in the first six weeks after kidding. The heavily milking doe is unable to take in sufficient food to meet her energy requirements and so uses body fat and protein reserves; this is the normal situation in early lactation. Metabolism of fat and protein by the liver leads to the formation of ketone bodies, small amounts of which can be utilised by muscles. If the appetite is further depressed or the energy deficit is great, ketone bodies build up in the body and can be detected in blood, milk and urine. The build-up of ketone bodies themselves further decreases the appetite and leads to the clinical signs of ketosis or acetonaemia, i.e. selective feeding (the goat eats hay but not concentrates), a drop in milk yield, a sweet smell of acetone on the breath and constipation.

Some goats after kidding have a more severe form of ketosis called *fatty liver syndrome*. Liver function is impaired by the large amount of fat deposited in the liver, so that the metabolism of fat and protein is affected and as the ketosis becomes more severe the goat stops eating completely, becomes thin and eventually collapses and dies. This condition occurs in overfat goats at kidding.

Treatment: Cases of acetonaemia usually respond well to treatment; fatty liver syndrome is often fatal even with veterinary treatment. Give 80 ml of glycerol (glycerine) in warm water daily for 3 or 4 days or 80–100 ml of propylene glycol, e.g. Ketol (Intervet).

Encourage the goat to eat; anything will do, e.g. browsings, evergreens such as ivy, cabbages etc. Stimulate appetite with products such as Restart (Bayer) and Rumen Stimulent (Univet).

Seek veterinary help if treatment does not produce a significant increase in appetite within 48 hours.

KIDS

Sensible husbandry procedures will give the kid a good start in life and help prevent infections. Preparations should start well before the birth of the kid. The pregnant goat should be given a booster vaccination (see page 106) and moved to the kidding area at least 14 days before kidding to enable her to manufacture the correct antibodies to deliver to the kid in the colostrum (first milk).

The kid should be born in a clean, well-strawed area with good ventilation, free from draughts. If necessary extra heat can be provided by means of an infra-red lamp. Thorough drying with a towel or straw stimulates the kid and prevents excessive heat loss.

Treat the kid's navel with an iodine preparation or antibiotic spray.

After birth, the kid rapidly uses its reserves of energy and dies unless these are replenished. All kids must receive colostrum within the first 6 hours of birth. Colostrum provides energy, warmth, fat soluble vitamins, has a laxative effect helping the kid to pass the meconium (first faeces) and most importantly provides antibodies which will protect the kid during its early life. If necessary, colostrum can be given by stomach tube (see page 131). Colostrum stores can be kept in the deep freeze in milk cartons or bags and thawed as necessary if a goat dies or has no colostrum for her kid.

Good hygiene should be maintained throughout the kid's life. Pens should be well strawed and not overcrowded, food and water containers should be free from faecal contamination and kids of different age groups should not be mixed together.

Routine Procedures

Disbudding
Horned kids should be disbudded by a veterinary surgeon between 2 and 7 days of age. Later than this, the horn buds are too large

to be easily dealt with by most disbudding irons. Most veterinary surgeons disbud kids under general anaesthesia and modern techniques mean that the procedure can be carried out from start to finish in about 15–20 minutes.

Castration

Male kids become sexually active at about 4 months of age and so should be castrated early if not required for breeding. Kids being reared for meat to about 5 months of age do not generally need castrating so long as they are separated from any females. Castration can be 'open' using a scalpel to remove the bottom of the scrotum and expose the testicles which are removed by a pulling action, or 'closed' using either a Burdizzo to crush the spermatic cord and blood vessels or a rubber ring which is a special ring placed above the testicles around the neck of the scrotum so restricting the blood supply. The testicles and scrotum wither and fall off in about 2 weeks.

Castration of kids over 2 months of age must be carried out by a veterinary surgeon.

Weak Kids

Weak kids can result from a number of causes such as premature birth, low birth weight, birth injury, malnutrition, congenital defects, exposure and infection.

Prematurity/low birth weight

Birth weights of kids are very variable ranging from single kids over 7 kg to kids of 2 kg in multiple biths. Birth weight is important as the larger the kid at birth the faster the growth rate.

Premature kids may have respiratory problems due to inadequate lung surfactant being produced, but kids up to 14 days premature have a good chance of survival if carefully looked after.

Birth injury

A difficult kidding can damage the kid by crushing the body as it passes through the pelvis, by over-enthusiastic pulling on limbs at an assisted kidding, or by asphyxia from pressure on the umbilical cord or reduced efficiency of the placenta during protracted uterine contractions. Birth injury can result directly in the death of a kid if the damage is severe enough, or indirectly from starvation or hypothermia by impairing feeding and movement, or it may result in a weak kid needing careful nursing and feeding.

Malnutrition before birth

Copper deficiency (enzootic ataxia, swayback). This can occur as a result of feeding roughage from copper deficient soils or from reduced copper utilisation, e.g. pastures top dressed with molybdenum and sulphur or heavy lime applications. It is a disease of grazing animals or animals fed grass-based diets and home-produced cereals. Zero grazed animals receiving a bought-in concentrate ration are unlikely to be affected.

Copper deficiency occurs in two forms; the congenital form where the kid is affected at birth or the delayed form where clinical signs do not appear until the kid is several weeks or even months old. In the congenital form, the kid is very weak and may die without showing obvious nervous signs. In the delayed form the kid is generally bright, willing to suck and eat but shows muscle tremors, head shaking and progressive weakness of the hind legs followed by paralysis. Other kids may show poor growth rates and increased susceptibility to infections and fractures. Similar signs will occur with poor nutrition, parasitism (see pages 106–110) or cobalt deficiency.

Treatment: Administer copper orally or by injection, but this is often unsuccessful except early in the disease.

Seek veterinary help whenever a deficiency is suspected. An accurate diagnosis is essential so that advice can be given to prevent further problems by supplementing does during early pregnancy. Overdosing with copper is dangerous.

Iodine deficiency (goitre). True iodine deficiency is probably rare in the UK. The condition is commonly misdiagnosed by goat-keepers when kids have any enlargement in the throat region. These swellings are usually enlargements of the thymus (see page 154).

Iodine deficiency arises from natural soil deficiencies of iodine or by over-feeding of goitrogenic feeds, e.g. kale and cabbage, which produces a secondary deficiency by stopping the thyroid gland accumulating iodine.

A lack of iodine causes abortion, stillbirths or very weak kids with thin hair cover, susceptible to cold and with breathing difficulties. The thyroid gland, slightly below and behind the larynx, is enlarged and this is known as a goitre. Less severely affected kids may be born very small and weak without an obvious goitre and other goats in the herd may have a reduced milk yield, poor appetite and poor growth rate.

Treatment: Supply older goats with iodised saltlicks, iodised salt or dose with potassium iodide. Affected kids can be treated with 3–5 drops of Lugol's iodine in milk daily for one week.

Veterinary help is essential if the condition is suspected to ensure an accurate diagnosis. An overdosage of iodine can itself cause disease.

Selenium deficiency (white muscle disease). A selenium/vitamin E deficiency affects the muscles, particularly of very young kids born to deficient dams. Kids may be affected from birth to 6 months of age, but generally it occurs between 2 and 16 weeks after birth with the most active kids being affected first.

If the heart muscles are affected the kid suddenly drops dead after or during exercise. If the skeletal muscles are affected the kid lies still and is sore with its leg muscles firm and painful on palpation.

Treatment: Inject the kid with a selenium/vitamin E preparation and prevent a deficiency by injecting the does and/or kids.

Veterinary help is essential as an overdosage of selenium will cause selenium poisoning.

Malnutrition after birth
Malnutrition will occur after birth if the kid is not able to suckle properly. The kid may have a physical defect such as a cleft palate or contracted tendons in the legs which prevent it from standing, or the mother may prevent the kid from feeding, she may have no milk, poor udder conformation or teat abnormalities. Where the dam is failing to provide sufficient milk the kid should be removed, stomach tubed and bottle fed as necessary.

Congenital defects
All kids should be inspected shortly after birth for obvious defects. Some will be incompatible with satisfactory growth and development and the kid should be humanely destroyed. These include having no anus (atresia ani), a cleft palate, hydrocephalus (water on the brain), missing or deformed limbs, spinal abnormalities etc. Others such as fish tail (double) teats and cryptochidism

(no testicles) or monorchidism (one testicle) are serious defects which are likely to be hereditary and again the kids should be culled. Supernumerary (extra) teats can easily be removed at the same time as disbudding but there is an ethical consideration here as the condition is likely to be hereditary and passed to future generations. Supernumerary teats should never be removed from male kids; these kids should be culled.

Kids should be rechecked at regular intervals during the first few weeks of life as supernumerary teats and fish tail teats may not be obvious at birth.

Contracted tendons, particularly of the forelimbs, are common in newborn kids resulting in an inability to straighten the leg. Mild cases with a partially bent leg will often resolve on their own as the tendons stretch with movement; more severe cases may need splinting to stretch the tendons and allow weight bearing on the foot. Kids with this condition will require help to suckle until the legs straighten.

Excessive motility of the limb joints is also common in newborn kids, but in most cases this corrects itself within a few days.

Hypothermia

Primary hypothermia results from direct exposure to cold, wet, windy weather when heat loss exceeds heat production. Small kids are more prone to hypothermia than large kids because they have a relatively large surface area relative to body weight and smaller energy reserves.

Secondary hypothermia occurs when newborn kids are unable to suckle and replenish their body reserves as a result of mismothering, birth injury etc.

The initial signs of hypothermia are shivering and aimless wandering followed by laying down quietly, coma and death. Treatment consists of drying the kid thoroughly with a towel, giving glucose by stomach tube or injection and placing the kid in a warm place such as by a radiator or fan heater.

Lamb reviver stomach tubes should form part of every goatkeeper's medicine cabinet. They consist of a plastic tube with a reservoir or syringe on the end. The tube is pushed gently down the throat over the tongue so that it enters the stomach (an easy technique when you have been shown how to do it). Glucose or colostrum can be fed at a rate of 50–75 ml/kg, i.e. 150–250 ml per feed or until the stomach feels full on palpation.

Experienced goatkeepers or veterinary surgeons can inject glucose directly into the abdomen. The kid is held dangling by its front legs and 10 ml/kg of a 20% glucose solution is injected 1 cm to the side and 2 cm behind the navel.

Infection

Congenital infections are those acquired before birth. Weak kids may be born as part of a herd problem involving stillbirths and abortions (see page 120).

In the first few hours after birth the kid is susceptible to infection from a number of organisms which gain entry via the navel or mouth resulting in infection of the navel, septicaemia, diarrhoea or joint-ill. Infection is most likely to occur in intensive kidding systems, particularly towards the end of the kidding period.

Joint-ill is an arthritis of young kids caused by bacterial infection of the joints. Infection occurs through the navel shortly after birth if the cord has not been adequately treated with iodine or antibiotic spray. The condition results in a sick, feverish kid with pain and swelling in one or many joints and it often responds poorly to treatment.

Failure of Kids to Grow Adequately

Possible causes of a poor growth rate are:

Nutrition

- Poor milk supply from the mother if the kid is being naturally reared, e.g. due to there being triplets, mastitis or inadequate feeding of the dam.
- Poor quality milk replacers in artificially reared herds or failure to establish regular feeding routines resulting in a poor intake of milk.
- Failure to feed a suitable concentrate ration to growing kids and over-reliance on poor quality pasture or hay.
- Mineral deficiencies—iodine (see page 129), copper (see page 129), selenium (see page 130) or cobalt.

Infections

- Gastrointestinal parasites (see page 106)
- Coccidiosis (see page 133)
- Severe lice infestation (see page 109)

Seek veterinary help early for kids which fail to thrive. An accurate diagnosis may involve tests on blood and/or faeces.

Diseases of Kids

Diarrhoea (scour)

Possible causes of diarrhoea
1. *Nutrition.* The majority of cases of diarrhoea in artificially reared kids are related to diet, either directly through sudden changes in the concentration or type of milk replacer, changing between goat milk and milk replacer, overfeeding or feeding milk at an incorrect temperature, or indirectly through dirty utensils or contaminated feed.

    In older kids, overfeeding of concentrates without adequate roughage either through bad management or stealing food will cause diarrhoea as will overgrazing lush grass, overfeeding greens or bread and mouldy feed.
2. *Infectious organisms.* A number of bacteria including E. coli, salmonella and Clostridia can produce severe diarrhoea in kids as can viruses such as Rotavirus and coronavirus, and protozoal organisms, particularly *Cryptosporidia* and *Coccidia*.

    Coccidiosis is a major cause of diarrhoea, particularly in housed kids. All goats are infected with coccidia and it is probable that all kids are infected during the first weeks of life by ingestion of contaminated food, bedding and water. Management standards determine whether or not the level of infection is sufficient to cause clinical signs of the disease which vary from mild diarrhoea, very sick kids with bloody diarrhoea and colic pains to sudden death. Coccidia are host specific and those of cattle, poultry or domestic pets do not affect goats.
3. *Parasitic gastroenteritis (roundworms).* Gastrointestinal parasitism is the major cause of diarrhoea in older kids. The worms cause damage to the stomach or intestines preventing proper absorption of digested food. Some species also suck blood. Lower numbers of worms produce unthriftiness and poor growth rates without obvious scouring, whilst very high numbers may result in sudden death. Many species are involved in scouring but the main ones are *Ostertagia* and *Trichostrongylus*. Goats share many of the same worms with cattle and sheep. The major blood suckers are *Ostertagia* and particularly *Haemonchus contortus*. These worms cause inflammation of the stomach, ulceration and a severe loss of blood resulting in anaemia and a loss of protein so that advanced cases develop water swellings (oedema) under the jaw (bottle jaw) and sometimes under the abdomen or legs.

Mixed infections produce scouring, anaemia and loss of condition.

Goats of all ages are susceptible but kids are often most severely affected. Adults may show signs of diarrhoea but infection may not be obvious and can result in weight loss and/or poor milk yield.

Goats become infected as they graze by eating the roundworm larvae. These pass into the stomach or intestines and develop into adult worms. Both stages can cause damage to the host animal. The adult worms produce large numbers of eggs which pass out with the goat's droppings onto the pasture. These hatch to produce larvae which can infect other goats.

Small numbers of larvae will not harm animals which are adequately fed. However, because development of the eggs and larvae on the pasture depends on the weather, particularly temperature and moisture (warm, moist conditions are ideal), large numbers of infective larvae can be ingested over a very short period, resulting in severe disease.

There are two possible sources of contamination for kids each year:

- *The pasture.* Most goats are kept on a very limited area, grazing the same paddocks year after year. Both worm eggs and larvae can survive on pasture over winter, so infection persists on permanent pasture from one year to the next, but eggs passed on the pasture in winter and early spring do not develop until the temperature in late spring is high enough. This leads to a peak of larvae on the pasture around midsummer and this is known as the 'spring rise' in larvae.

  The dangers of larvae on pasture can be reduced by resting paddocks and grazing fodder crops such as kale or lucerne, and leaving the pasture ungrazed until mid-July when the overwintered larvae will have died off.

- *Older goats.* In the autumn ingested larvae are inhibited and remain in a dormant state within the goat until the following spring when they resume development, mature and pass eggs onto the pasture. There is an additional rise in the number of eggs around kidding because of a temporary relaxation of immunity. This is known as the 'periparturient rise' in larvae.

  Thus nature increases the worm burden on the pasture in the spring when conditions are right for development and new hosts, the kids, are available for infection.

  If clean, worm-free grazing is not available, control relies on the sensible use of drugs known as anthelmintics given at strategic times throughout the year (see page 107).

Preventing diarrhoea

1. *The dam.* A booster vaccination against clostridial diseases should be given to the dam 3–4 weeks before kidding. She should be placed in the kidding area at least 14 days before kidding to have time to produce the correct antibodies to pass on in her colostrum.
2. *The kid.* All kids should receive adequate colostrum (at least 200 ml and preferably more), during the first 12 hours after birth.
3. *The environment.* Avoid overcrowding. Provide clean, dry, well-strawed pens for each batch of kids, and do not mix kids of different age groups. Place food and water containers above the floor to avoid faecal contamination.
4. *Feeding.* Ensure that milk or milk replacer is fed at the right concentration in the right amounts, at the correct temperature and at regular times.

   All feeding utensils should be clean and bottles and bowls for milk feeding should be sterilised.
5. *Anthelmintic treatment.* Kids should be treated regularly with anthelmintics throughout the grazing season (see page 107) unless they are being very extensively grazed or can be moved regularly to fresh pasture.

Treatment: Take unweaned kids off milk for 24 hours. Feed warm water with 1 oz of glucose/litre or an electrolyte (salt) replacer which can be obtained from your veterinary surgeon; then feed half the normal quantity of milk for the next 2 or 3 days. Do not feed milk replacers at lower than the recommended concentration or dilute whole milk; only milk of the correct strength will clot normally in the stomach.

Restrict solid food, particularly concentrates, for 24 hours, gradually increasing to the normal amount as the diarrhoea is controlled; correct any obvious overfeeding and review feeding practices (see chapter 4). Give a kaolin suspension, 2–10 ml three times daily and anthelmintic treatment to any grazing kids; avoid underdosing (weigh the kid if possible).

Seek veterinary help if; the diarrhoea persists despite treatment; the kid shows persistent abdominal pain or bloat; the diarrhoea is profuse or blood is present; the kid is very depressed.

Colic (abdominal pain)

Colic is the term used to describe the symptoms produced by abdominal pain.

The goat shows pain in several ways: lethargy, grinding its teeth, yawning, constant stretching of the back, failure to drink milk, pawing the ground, rapid breathing, and in severe cases a kid may scream with pain and throw itself around. Other signs such as distension of the abdomen by gas (bloat) in the true stomach, rumen or intestines may become apparent.

Colic signs in kids should be treated as a potential emergency as kids may die after quite short periods of abdominal pain. If in doubt seek veterinary help immediately. All kids should be checked routinely after feeding and particularly after the last feed at night.

Possible causes of colic

- *Diarrhoea* (see page 133)
- *Excessive fermentation of milk* following a change to milk replacer, or overfeeding, feeding replacer at the wrong concentration or temperature etc. which can lead to gas build-up in the true stomach (abomasum) and signs of pain from the resulting abomasal bloat
- *Rumen bloat* in older kids can occur during weaning if milk enters the rumen rather than the true stomach, if movement of the stomach is inhibited or if an obstruction prevents the release of gas
- *Constipation*
- *Enterotoxaemia* (see page 104)
- *Plant poisoning* (see page 143)
- *Urinary calculi* (bladder stones) can block the penis in older male goats (see page 137).

Treatment: Check the kid at regular intervals, particularly if the pain persists or worsens or if bloat is present.

Mild cases of bloat will respond to a drench of a tablespoonful (15 ml) of vegetable oil with a teaspoonful of sodium bicarbonate, or a proprietary bloat drench. Seek immediate veterinary help if the pain is severe or breathing is difficult.

To treat constipation drench with a tablespoonful (15 ml) of liquid paraffin with about 2 teaspoonfuls (10 ml) of vegetable oil, or a proprietary colic drench containing turpentine oil and polymethylisoloxone, e.g. Gaseous Fluid (Day, Son & Hewitt).

For diarrhoea see page 133.

Urinary calculi (bladder stones)

Urinary calculi or bladder stones are a problem of male goats from the age of 4 months upwards. The condition is caused by the deposit of phosphate crystals, usually as a result of feeding high concentrate/low fibre diets or because of low voluntary water intake. In the bladder the stones are usually harmless but small stones passed in the urine get lodged in the channel through the penis, the urethra, blocking the passage of any more urine. Because the urethra is narrower in castrated animals, they are more likely to be affected.

The affected goat shows increasing signs of colic (see page 136), with frequent unsuccessful attempts to urinate. Examination of the penis may reveal crystals on the surrounding hairs.

To prevent urinary calculi all male goats should have clean water available 24 hours a day. The water should be changed twice daily. They should also have plenty of palatable roughage offered to them, e.g. good hay.

Seek veterinary help immediately the condition is suspected.

CONDITIONS OF THE UDDER AND TEATS

Cuts

Prompt attention to injuries of the udder and teats is essential to prevent infection entering the udder and to ensure rapid healing, particularly of teats where even small wounds can be repeatedly traumatised by milking.

Treatment: Superficial wounds should be cleaned with antiseptic solutions, antibiotic sprays or wound powder. Small wounds can be closed with adhesive tape or spray-on human skin preparations. Particular attention should be paid to injuries near the teat opening as even small cuts can lead to gangrenous mastitis.

Seek veterinary help for deeper wounds with flaps of skin or penetration into the teat with leakage of milk.

Bruising

Trauma to the udder and teats, often by headbutting and being trodden on, can produce painful, hard swellings which need to be distinguished from inflammation caused by mastitis.

Treatment: Cold water or ice applied 3 or 4 times daily will control swelling and pain. After several days of cold treatment, gentle massage and hot compresses will promote healing.

Lumps in the Udder

Abscesses

Abscesses in the udder can arise following episodes of mastitis (see page 139) or from penetrating wounds. Deep abscesses are not possible to treat satisfactorily but superficial abscesses can be drained.

Fibrous scar tissue

Fibrous lumps arise as a sequel to mastitis (see page 139) or trauma. They are an obvious fault in the show ring so numerous methods are adopted to try and get rid of them. Remember, most new lumps will gradually reduce in size through natural healing processes with no treatment necessary. This is why there are so many different 'cures' for the condition. Topical applications—anti-inflammatory creams, herbal remedies etc.—will reduce the size of surface lumps, but the only successful treatment I have found for deeper lumps is cold laser therapy (many equine veterinary surgeons and large equine establishments have these machines).

Pustular Dermatitis of the Udder (Goat Pox)

Pustular dermatitis of the udder, colloquially known as goat pox, although true goat pox is not present in the UK, is caused by a bacterial infection with *Staphylococus aureus*. Pustules of varying size appear on the udder, teats and groin and occasionally the ventral abdomen or other parts of the body. The infection enters the skin when the udder is traumatised, e.g. by prickly straw, and is most common in first kidders soon after parturition as their udders are more tender. The condition is more severe in warm weather, and the pustules are easily broken and spread, healing as scabby lesions. The lesions are not generally painful and milk yield is not affected. Infection is easily spread between goats at milking by hands, cloths or milking clusters.

Treatment: Use antiseptic washes, such as savlon or hibiscrub, or good lanolin-based udder creams, (avoid cetrimide-based udder creams which dry the skin). Milk affected goats last, paying particular attention to hygiene between goats (see pages 56–57); use a teat dip.

Seek veterinary attention if the lesions persist as these may require antibiotic treatment.

Fly Bites

Biting flies may produce quite severe lesions, superficially resembling pustular dermatitis but generally more swollen and painful.

Treatment: As for pustular dermatitis. Control flies in the goat house with preparations such as Golden Malorin (Sanoffi).

Mastitis

Normally the inside of the udder and the milk produced is sterile. When bacteria or other micro-organisms enter the udder, mastitis occurs. Mastitis literally means inflammation of the udder and results in physiological, chemical and bacteriological changes in the milk and changes in the udder itself.

Various factors predispose the goat to mastitis. These include physical injury (butting by other goats, treading on teats, cuts on the udder or teats), over-milking, trauma caused by poor hand-milking technique or faulty milking machines, poor hygiene at milking, dirty bedding and the presence of infection in other goats in the herd.

Because mastitis is caused by micro-organisms it can easily be spread from goat to goat, e.g. by hands when handmilking, by the clusters of milking machines or by udder cloths. For this reason, any known cases of mastitis should always be milked last and any cloths used to clean udders should be disposable (preferably paper) and never used for more than one goat.

Mild clinical mastitis

Most cases of mastitis noticed by goatkeepers are fairly mild. Generally the goat herself is not ill but changes are obvious in the milk; poor keeping quality, curdling on boiling, milk taint or clots in the milk.

N.B. Any of these changes may indicated mastitis but some may be due to other problems. Thus poor keeping quality may be due to poor hygiene, curdling on boiling may be due to high levels of colostrum persisting for some time after kidding, and milk taint can arise from a number of causes (see page 142).

Acute mastitis
Here the goat is ill—lethargic, feverish, off its food—and the udder is hard, swollen and painful. Milk yield falls and the consistency of the milk is changed; often it is thin and watery with clots.

Peracute mastitis (gangrenous mastitis)
The goat is usually very ill (or even found dead), initially with a high fever, although later as she becomes more shocked, the temperature may become subnormal. The udder is very hard, hot and swollen, later turning blue as gangrene sets in. There is no milk, only a thin, bloody secretion from the teat. Many goats die from toxaemia. In those that survive the blood supply to the infected half of the udder is lost and the tissue eventually sloughs off.

Chronic mastitis
Repeat episodes of mild mastitis will lead to chronic changes in the udder, notably fibrosis and atrophy. Fibrosis produces persistent lumps and hardness of the udder tissue and atrophy means that functional milk-producing cells are lost and the udder shrinks.

Subclinical mastitis
This is the most insidious form of the disease and, although very common, it often goes undetected because the udder and milk appear normal. There is usually some reduction in milk yield and there may be a reduction in the levels of butterfat and solids-not-fat and a decrease in keeping quality. The true level of subclinical mastitis is unknown but 4–6% of udder halves may have bacterial infections. Subclinical mastitis is only detected by examining a milk sample.

Seek veterinary advice whenever mastitis is suspected. Even mild cases may progress to a more severe form or the goat may become chronically infected. If possible submit a milk sample so that the bacteria causing the infection can be identified and its drug-sensitivity determined.

Collect milk samples carefully. Clean the teats thoroughly, wipe the teat end with alcohol or spirit, discard the foremilk and collect

20 ml of milk from each half of the udder in a sterile container. Any milk which cannot be tested immediately should be stored at 4°C or frozen.

Treatment: Apply general support for a sick goat (see page 112) and give antibiotics by intramuscular injection or intramammary tubes. At present there are no intramammary preparations licensed for goats in the UK. This means that milk from treated animals should not be used for human consumption for a minimum of 7 days after the insertion of the last tube. Care must be taken when inserting tubes into the teat; the ends of the teats should be cleaned with alcohol or spirit before insertion. Some teat nozzles are quite large and it is easy to traumatise the teat or even introduce further infection. Use one tube per side of the udder; *never* attempt to split the tube between the two halves. The frequency of insertions of the tubes and the number of tubes required varies with different preparations. The affected half of the udder should be completely stripped out before inserting the tube and the teat then immersed in a teat dip.

Dry goat therapy is the insertion of a long-acting antibiotic preparation into the teat at drying off. It is only necessary in goats which have had an episode of mastitis during the previous lactation or if subclinical mastitis is suspected.

'Pink Milk' (Blood in the Milk)

Bleeding into the udder from a damaged blood vessel commonly occurs in first kidders in the first few days after kidding. It may also occur at other times, particularly if the goat is milked by someone other than the usual goatkeeper. Sometimes actual pink milk is seen or there may be a pink sediment after the milk has been allowed to stand. The condition is usually self-limiting after a few days and needs no treatment.

Seek veterinary help only if you are worried about mastitis. Culture of a sterile milk sample will confirm the absence of infection. Bloody milk is not a common sign of mastitis.

Udder Oedema

Some does develop very swollen udders just before kidding. This is a normal physiological process and in virtually all cases the swelling disappears over a few days as milking commences.

Seek veterinary help if the problem persists, if the udder is hard and hotter than normal (mastitis may be present), or if little or no milk is produced (this may be due to 'hard udder', see below).

'Hard Udder'

Goats infected with the CAE virus (see page 111) may develop very firm, swollen udders at kidding. Unlike goats with udder oedema, little milk is produced and milk let-down is poor. Softening of the udder occurs after about a week but milk production remains below normal.

A Fall In Milk Yield

This is a common occurrence in many illnesses, particularly where the goat is feverish or off her food.

Seek veterinary help if the milk yield does not improve over 48 hours and there is no obvious reason, such as bullying, stress etc., for the drop in milk.

Milk Taint

Milk taint may occur as a herd or individual goat problem. If it is a herd problem there are several possible causes:

- *Feed taint.* Certain foods such as kale, turnips, swedes, garlic and cow parsley flavour milk. Any possible sources of taint should be identified and eliminated. If it is necessary to feed them, they should be fed immediately after milking.

  Housing the animals and feeding hay will eliminate the grazing as a source of the problem.
- *Genetic factors.* Certain families of goats have a high milk enzyme activity leading to the release of 'goaty-tasting' fatty acids.
- *A vitamin B_{12} deficiency* will result in taints. This can be produced by cobalt deficiency or by gastrointestinal worms.
- *Mastitis.* Widespread subclinical mastitis may cause a herd taint problem.
- *A poor milking technique* can lead to bacterial contamination of milk.
- *Poor dairy hygiene,* e.g. inadequate filtration, cooling, refrigeration or freezing of milk, can also lead to bacterial contamination.

- *Agitation of milk*, e.g. in the milking machine line, can cause the release of 'goaty' fatty acids.
- *Exposure of the milk to copper or iron* gives an oily/cardboard flavour.
- *Exposure of the milk to sunlight or fluorescent light* gives a flat/burnt flavour.
- *Storage taint*. Milk readily picks up strong flavours from other foods.

An individual goat problem can result from any of the above although of course if the goat is with a large number of animals which are not affected, many of the possibilities can be eliminated. The problem goat should first be identified by checking each goat's milk and then each half of the udder should be checked individually; where only one half of the udder is affected, mastitis is usually the problem. If both halves are affected, mastitis may still be the problem and the udder should be checked carefully for lumps, fibrosis, abscesses etc. Milk samples can be taken from individual halves and submitted to your veterinary surgeon for bacteriological testing.

PLANT POISONING

Goats, being inquisitive and browsing animals, commonly consume potentially poisonous plants. As with all diseases, prevention is better than cure and any poisonous plants in the vicinity should be identified and if possible removed. It is safest to assume that *all* evergreen shrubs are poisonous, unless you definitely know they are safe. Most cases of poisoning are caused by garden shrubs, in particular, rhododendrons, azaleas and laurels, often from trimmings discarded in neighbouring gardens.

Some plants are harmless when fresh but poisonous when dry and wilted, e.g. leaves of the Prunus family (cherry, plum etc.). Certain plants are equally toxic if eaten fresh or dried in hay (e.g. ragwort). Some cause a delayed poisoning, e.g. ragwort, St John's wort.

Poisoning may not always be obvious due to the variety of symptoms, though some such as vomiting are nearly always due to poisoning.

The Clinical Signs and Plants Involved in Poisoning

Anaemia

Kale
Rape

Constipation

Linseed
Oak (acorns and old leaves)
Ragwort

Diarrhoea

Box
Castor seed (in foodstuffs)
Foxglove
Hemlock
Linseed
Oak (young leaves)
Potato (green)
Rhododendron
Water dropwort
Wild arum

Discoloured urine

Bracken
Brussels sprouts
Cabbage
Kale
Oak
Rape

Frothy bloat

Legumes, e.g. rapidly
    growing lucerne

Goitre and stillbirth

Brassica spp.
Linseed
Some clovers

Haemorrhage

Bracken

Nervous Signs

Black nightshade
Common sorrel
Hemlock
Horsetails
Laburnum
Male fern
Potato
Prunus family
Ragwort
Rape
Rhododendron
Rhubarb
Sugar beet tops
Water dropwort

Oestrus

Some clovers

Photosensitization

Buckwheat
Bog asphodel
Ragwort
St John's wort

Stomatitis

Cuckoo pint
Giant hogweed

Sudden death

Foxglove
Laurel
Linseed
Water dropwort
Yew

Vomiting

Azalea
Rhododendron

What to do if poisoning is suspected

1. *Do not panic.* Goats are often able to consume small quantities of poisonous plants without ill effect, particularly when the rumen is full of other foodstuffs. Thus, for instance, moderate quantities of rhubarb and green potatoes can be safely dealt with by the goat whereas large quantities would prove toxic.
2. *Separate the goat from the plant.* It may be possible to actually remove the plant material from the goat's mouth. Identify the plant where possible.
3. *Keep the goat walking slowly* so that it does not settle and start cudding.
4. If the goat is shocked put her in a warm place and rug her.
5. Administer strong *tea* in large quantities (do not attempt to dose an animal which is vomiting). The tannic acid in tea

precipitates many plant alkaloids and salts of heavy metals and has a useful stimulant effect. Strong *coffee* also has a stimulant effect.

6. Give *liquid paraffin* for its mild laxative action. Kids need 5–10 ml, adults 150–300 ml.
7. Give a mixture of eggs, sugar and milk as a *demulcent* to soothe and relieve irritation of the stomach lining.

Seek veterinary help if symptoms of poisoning persist. Specific antidotes are available for certain poisons and the removal of rumen contents by rumenostomy may be indicated. In other cases antibiotics, vitamins, fluid therapy or treatment for shock will be necessary.

SKIN PROBLEMS

Goats may be affected by a number of skin diseases some of which cause intense irritation while others are largely non-irritant.

Irritant Skin Problems

Lice
Lice are the commonest cause of skin irritation particularly during the winter. Because lice are visible to the naked eye, this condition is easily diagnosed and treated (see page 109).

Flies
Fly bites in the summer also cause irritation. The udder is affected most and may be covered with pustules and raised, red, sore areas. These lesions are similar to those of pustular dermatitis (see page 138) but are generally more irritating. Antiseptic and anti-inflammatory creams and udder washes will soothe the sore areas. Flies in the goat house can be controlled with preparations such as Golden Malrin (Sanoffi).

Sarcoptic mange
The disease is relatively common in goats of all ages and is caused by a mite which burrows into the skin. Hair loss starts around the eyes, ears and nose with reddening and a few raised areas. As the irritation increases there is progressive hair loss with thickening and wrinkling of the skin as the goat scratches more and more.

Eventually the whole head, neck and body may be involved. Treatment is difficult and veterinary help is needed.

Other external parasites such as harvest mites, forage mites, cheyletiella mites, fleas and poultry mites occasionally infest goats.

Scrapie

Scrapie is a progressive fatal disease of the central nervous system of sheep and goats. Most animals are probably infected shortly after birth, but because the disease has a long incubation period it is not usually seen until the goat is 3 or more years old. Two forms of the disease have been described, pruritic and nervous, but many animals show a combination of intense itching—scratching with the hind feet, rubbing the head and back, rubbing the skin or even biting—and nervous signs ranging from slight behavioural changes to severe incoordination and eventually coma and death. There is no treatment. Scrapie is now a notifiable disease and must be reported to the Ministry of Agriculture if suspected.

Non-irritant Skin Problems

Chorioptic mange (stable mange, leg mange)

Caused by a burrowing mite, this is a very common condition of goats in the UK particularly during the winter. Goats should always be checked during routine foot trimming; white/brown scabby lesions start at the back of the heel and extend up to the pasterns or even the knee or hock. In severe infections, the abdomen, sternum or even the upper body may be affected. When picked off, the scabs reveal raw, sore areas. Treat by thorough washing with a suitable insecticidal shampoo (wearing gloves). The legs should be well soaked and as much scabbing as possible removed. Severe infestations will require several washes, and localised areas can be treated daily. Some cases will require antibiotic injections and sprays to combat secondary bacterial infection.

Demodectic mange

This is relatively common, particularly in British Alpine goats, and goatlings commonly show clinical disease following infection as a kid. The demodectic mange mite multiplies in the sebaceous glands of the skin producing small nodules or pustules on the head, neck and body. Yellow, pussy material can be squeezed from the nodule. Veterinary assistance is required to confirm the diagnosis and for treatment.

Urine scald

When male goats urinate on themselves during the breeding season, particularly down the back of the forelegs, face and beard, this causes staining, hair loss and inflammation of the skin. Similar staining occurs in goats that are recumbent or housed in dirty conditions. Older male goats commonly have thickened scaley skin over the head and back. Regular shampooing will help prevent the condition developing. Body oils or olive oil thinned with surgical spirit has been used to remove excess scaling and allow new hair to grow.

Ringworm

This is an uncommon fungal infection in goats which is acquired from other animals such as cattle, rodents, dogs and cats. Grey-white, raised, crusty lesions, initially circular but later irregular in shape, occur. Veterinary help is needed to confirm the diagnosis and for treatment.

Certain skin diseases only occur in particular breeds of goats. In *Pygmy Goats Syndrome*, affected animals develop crusty lesions around the eyes, ears, nose, head and tail. The *Golden Guernsey Goat Syndrome* is an hereditary condition which results in sticky kids born with greasy, matted coats which remain abnormal throughout life.

LAMENESS

In kids, lameness is most commonly due to an accident or trauma resulting in bruising, sprains, strains or fractures particularly of the front legs.

In adult goats, although accidents are common, most lamenesses are caused by poor management or poor foot care.

Infectious Diseases of the Foot

Interdigital dermatitis (scald)

Continuous wetting of the foot and interdigital skin damages the tissues and allows the invasion of the bacterium *Fusobacterium necrophorus* resulting in inflammation and swelling of the skin between the hooves. Affected animals are lame in one or more feet and may walk on their knees. Scald often spreads rapidly throughout the herd.

Uncomplicated cases of interdigital dermatitis recover quickly if the animals are moved to dry conditions and the feet treated with antibiotic spray or run through a footbath.

Footrot

Whereas *Fusobacterium necrophorus* does not damage the horn itself, another bacterium, *Bacteroides nodosus*, can act with it to invade the hoof and sole and cause benign footrot or virulent footrot depending on the strain of *Bacteroides* present. There is separation and undermining of the sole with a pussy discharge, and the horn can be pared away to show greyish soft horn with a characteristic foul smell.

All the feet should be trimmed to remove all infected and under-run tissue. The shears should be disinfected between goats and all foot parings disposed of carefully. Individual goats can be treated with antibiotics and antibiotic sprays. With a herd problem the goats should be run through a footbath weekly. Zinc sulphate 10% is probably the best solution to use; add 10 kg of zinc sulphate to 100 litres of water. Goats should be penned on dry ground until their feet are dry.

Puncture of the sole

Penetration of the sole by a sharp object such as a nail leads to abscess formation with pain and lameness until the pus is released by paring and the area thoroughly cleaned.

More severe *foot abscesses* can occur following footrot or a deep penetration wound with the abscess bursting out at the coronet or interdigital space.

Foot and mouth disease

This is not normally present in the UK but goatkeepers should be aware of it as it is a notifiable disease and must be reported to the police or Ministry of Agriculture if suspected. It produces blisters between the claws and at the coronet, resulting in acute lameness, generally in all four feet. There may also be small blisters on the lips, tongue and teats.

Non-infectious Diseases of the Foot

White line disease

In wet conditions the horn of the wall of the foot will separate from the sole along the white line leaving a space which fills with dirt and debris and causes lameness. If infection occurs in this area, pus will collect and may develop into an abscess.

Foreign bodies
Foreign bodies such as stones or nails may become embedded in the foot or interdigital area.

Laminitis
Laminitis is an inflammation of the sensitive tissues of the hoof beneath the horn.

Acute laminitis
This may occur after any toxic condition such as mastitis, metritis, or retained foetal membranes. It is also common a few days after kidding, with or without any of these conditions, after over-eating concentrates, in silage-fed goats because of continued ingestion of acid silage and in female goats fed a high energy, high protein ration. It also occurs in male goats who are overfed during the summer.

The goat suddenly appears lame, generally on both front feet but occasionally on all four feet. There is a disinclination to walk, the goat prefering to lie down and walk on its knees or shift its weight from foot to foot to try to relieve the pain. Other signs of pain such as teeth grinding may occur and the goat is feverish, inappetent and its milk yield drops. The toes of the affected feet feel cold and the coronet region warm.

Subclinical laminitis
This is a common disease of dairy goats fed large amounts of concentrates. Although obvious signs of acute laminitis are not present, changes in the laminae lead to the development of hoof abnormalities such as uneven claw growth. Bleeding occurs into the wall, heel and sole and is evident at foot trimming as fine reddish streaks in the horn.

Chronic laminitis
Where acute laminitis is not recognised or treated because horn formation is disturbed, chronic laminitis develops leading to overlong feet ('sledge runner foot') or very high feet ('platform soles'). Anglo Nubian goats in particular develop very hard, high soles, although the general shape of the foot remains fairly normal. These goats have a characteristic goose-stepping walk and spend a lot of time on their knees.

Most cases of laminitis can be controlled by better management, particularly correct feeding. Acute cases should be fed a reduced protein/energy diet; antibiotics should be given to control any infections or toxic cause, together with painkillers. Heat treatment of the feet will help restore the circulation if used during the first few hours of infection; cold water is of benefit later and for about 7 days. Chronic cases need careful foot trimming to relieve pain and regular, repeated foot care when the foot is grossly over-grown or misshapen.

EYE INFECTIONS

Conjunctivitis

Damage or inflammation of the eyes results in conjunctivitis with reddening of the eyelids and excessive tear production. The affected eye may be kept closed and the goat may avoid light. In more severe cases, the cornea (window) of the eye loses its opacity, becomes cloudy and looks a whitish-blue colour. Even with treatment the cloudiness of the cornea may persist for several weeks.

If only one eye of an individual goat is affected, the cause is likely to be trauma, e.g. poking a hay stalk or twig into the eye or the presence of a foreign body, such as a piece of grass or shaving, behind the third eyelid.

Pink eye, (infectious Keratoconjunctivitis) is caused by *Mycoplasma conjunctivae* and a number of other organisms. It is a very infectious disease transmitted by contact with carrier goats at shows or by introducing a new goat into the herd. Flies and lice may help to transmit it.

Kids may develop sore eyes if they have inturning eyelids (**entropion**) but the congenital condition is rare in the UK.

Veterinary help should be sought whenever the conjunctivitis is severe, if more than one goat is affected, if a foreign body cannot be easily removed or if entropion is suspected.

RESPIRATORY DISEASES

A variety of respiratory problems affect goats, ranging from very serious pneumonia to slight coughs and runny noses.

Nasal Discharges ('Colds')

Infection with viruses, bacteria or fungi or dusty conditions can give rise to colds. Mild colds will usually resolve themselves without treatment but if the animal goes off its food, veterinary advice should be sought. If the discharge is one-sided, it may mean that there is a foreign body, such as a grass seed, up the nose and the veterinary surgeon should be consulted.

Coughs

Coughs are common, particularly among show animals, and may persist for some time even with treatment. As with colds, treatment is only necessary if the animal appears ill. Veterinary advice should be sought for any severe sudden coughing in a sick animal as it may indicate the presence of pneumonia (see page 151), heart failure, oesophageal obstruction or the presence of lungworms (see page 152).

Allergic bronchitis occasionally causes a persistent cough when a goat becomes sensitive to hay or straw dust. Sensitive animals should be bedded on shavings and fed dried lucerne or haylage rather than hay.

Difficulty In Breathing

Apparent difficulty in breathing—rapid respirations, mouth breathing, panting—may be caused by a primary respiratory problem such as pneumonia but can also be caused by other conditions such as anaemia, bloat, hypocalcaemia or heat stroke. Unless there is an obvious cause, e.g. heat stroke, veterinary help is essential to arrive at a correct diagnosis and start the correct treatment.

Pneumonia
Pneumonia means infection and inflammation of the lung.

- *Bacterial pneumonia.* A number of different bacteria can cause pneumonia but the most important is *Pasteurella haemolytica* which causes pasteurellosis. This is a disease which strikes very rapidly; animals may be found dead or running a high temperature, with rapid breathing and a dry cough. It may be precipitated by stress, e.g. transport, moving etc. Vaccination against pasteurellosis is available using sheep vaccines.
- *Inhalation pneumonia.* This may follow the inhalation of fluids

into the windpipe and lungs during drenching, force feeding of kids from a bucket or after rhododendron poisoning.

- *Parasitic pneumonia (lungworm infection).* Goats can be infected by two types of lungworm: *Dictyocaulus filaria* and *Muellerius capillaris.* Symptoms are coughing, particularly after exercise, and failure to thrive. More severe illness may occur with rapid breathing and discharges from the eyes and nose particularly if a secondary bacterial infection occurs. Control methods for gastrointestinal parasites (see page 107) also apply to lungworms, but not all anthelmintics are effective.

Heat stroke

This is common in housed goats in the summer, in goats being transported to shows or goats tethered in the sun without shelter. The goat breathes rapidly and pants loudly with an open mouth in an attempt to lose body heat and the rectal temperature may reach 41.5°C (107°F) or more. The best method of cooling is a thorough dousing with cold water which should rapidly reduce the temperature.

Bloat

Bloat occurs whenever gas is unable to escape from the rumen. As gas builds up, the left flank of the goat becomes distended by the large gas-filled rumen and eventually breathing is difficult because of the pressure on the diaphragm. Bloat can arise due to: an obstruction of the oesophagus ('choke'), where a piece of apple or sugar beet pulp etc. becomes stuck preventing the escape of rumen gas; paralysis, e.g. due to tetanus; the excessive production of gas from eating large amounts of fermentable foods; or as a result of grazing lush pasture such as clover or lucerne which produces a large amount of frothy material. This is known as 'frothy bloat'.

First aid treatment consists of drenching with 50–100 ml of vegetable oil or a proprietary bloat drench. In an emergency, 10 ml of washing-up liquid will help dispense frothy bloat. Massage the abdomen to spread the oil. Gentle exercise encourages the goat to belch. Increase the fibre in the diet by feeding hay before a return to feeding clover or grazing.

Any case of bloat which does not respond rapidly to first aid measures requires veterinary treatment. If necessary gas can be released by stomach tube or by inserting a large needle or trochar and cannula into the rumen through the left flank.

Control of Infectious Respiratory Disease

Strong, healthy kids are less susceptible to respiratory infection. Avoid mixing different age groups in the same air space; operate an all in/all out policy for batches of kids. All goats should be housed in well ventilated, draught-free buildings. Any affected animals should be isolated and the others carefully observed for early signs of disease.

EXTERNAL LUMPS

Skin tumours are uncommon in the UK. Papillomas, squamous cell carcinomas and malignant melanomas occasionally occur, particularly around the head and neck, on the udder or around the vulva and anus. Other tumours such as lymphosarcoma involve the enlargement of lymph nodes particularly in the face and shoulder region.

Abscesses occur occasionally, particularly around the face, as a result of a small penetration wound becoming infected with bacteria.

Caseous lymphadenitis is common outside the UK and results in abscesses in the head and neck region and many other lymph nodes throughout the body.

Vaccine reactions commonly occur at the clostridial vaccine injection sites (see page 106).

Orf is a virus infection causing pustules and then crusty, scabby lesions on the lips, gums, nostrils and occasionally the udder, feet or tail.

A number of swellings occur in the neck region of kids:

Salivary cysts are relatively common in Anglo Nubian goats and occur occasionally in other breeds. Soft fluid swellings form just below the jaw and may become quite large and need surgical removal.

Thymic hyperplasia results in soft but solid swellings in the ventral neck region. This is probably a normal development occurrence which resolves itself, but is often confused by goat-keepers with 'goitre' (see below).

Thyroid enlargement ('goitre') involves swellings either side of the windpipe, slightly below or behind the larynx (see page 129).

Wattle cysts are swellings at the base of one or both wattles varying from pea size to several centimetres in diameter. They occur particularly in British Alpine and Anglo Nubian goats.

FURTHER READING

The Goatkeeper's Veterinary Book, Peter Dunn, published by Farming Press Ltd, 3rd edition, 1994.
Outline of Clinical Diagnosis in the Goat, John Matthews, published by Wright, 1991.

APPENDICES

APPENDIX 1

Buyers' Guide

Books on goat subjects	British Goat Society Secretary 34–36 Fore Street, Bovey Tracey, Newton Abbot, Devon TQ13 9AD Tel: 01626 833168
Packaging of milk, cheese, yoghourt, etc.	Goat Nutrition Ltd., Biddenden, Ashford, Kent TN27 8BL. Tel: 01580 291548
Dairy articles, churns, pails, coolers, separators, cheese and yoghourt starters, health and veterinary products	Lincolnshire Smallholders Supplies Ltd., Willow Farm, Thorpe Fendyke, Wainfleet, Lincs PE24 4QE
Vitamins (Caprivite), kids' milk replacer (Capriolac)	Goat Nutrition Ltd., Biddenden, Ashford, Kent TN27 8BL
Wormers (Panacur)	Hoechst Animal Health UK Ltd., Walton Manor Road, Walton, Milton Keynes, Bucks. MK7 7AJ
Lamb and kid feeders—multi teated	AK Enterprises, Blakemoor Farm, Longhope, Little London, Glos GL17 OPH
Mineral blocks (Rockies)	Tithebarn Products Ltd., Southport, Lancs
Dried lucerne pellets; natural choice range of feeds	Dengie Crops Ltd., Southminster, Essex CMO 7JF. Tel: 01621 773883
Seeds: herbal strip mixtures, farm seeds in small quantities, kales, mangolds, comfrey	SM Mcard (GT), 39 West Road, Pointon, Sleaford, Lincs NG34 ONA

Milking stands, hay racks, portable racks, paths	BH Cox Fabrications, Slately Hall Farmhouse, Trinity Road, Kingsbury, Tamworth, Staffs B78 3EW
Electric Fencing	Flexinet, Unit C, Chancel Close, Trading Estate, Estate Avenue, Glos G44 7SH
Fencing aids	Drivall Ltd., Narrow Lane, Halesowen, West Midlands B62 6PA
Pasteurisers	Goat Nutrition Ltd., Biddenden, Ashford, Kent TN27 8BL
Packaging of cheese, yoghourt, etc. pots and lids	Cockxsudbury Ltd., Unit 9, Alexandra Road, Sudbury, Suffolk
Goathage, baled lucerne pre-packed in covers	Lawrence Buchanan, Manor Farm, Winstone, Cirencester GL7 7JU
Low cost housing: adaptable polypen	Polypen Animal Housing, Unit 9, Tewkesbury End Centre, Delta Drive, Tewkesbury, Glos GL20 8HR
Housing	Pentangle Animal Housing, Pentangle European. (Not for 'hobby' goatkeeping.) Tel: 01404 83420/46416
Bacterial testing labs	E. W. Love, B.Sc., PhD., Labitest, Unit 25, Bramwell Workshops, Barnwell, Oundle, Peterborough, Cambs PE8 5PP
Fibre evaluation	E. C. M. Downing & E. J. B. Downing, Hainford Place, Hainford, Norwich, Norfolk NR10 3BX
Export advice	Agricultural Export Services, Pool House, Birdlip, Glos

Various Ministry of Agriculture, Fisheries and Food publications including labelling, packaging, cheese bulletins, milk regulations, poisonous plants and guide lines on 'use by' dates	See MAFF Publications Catalogue, listing HMSO/MAFF titles, tel: 0181 694 8862 or 0171 873 0011. Helpline number: 01645 335 577
Insurance for goats: comprehensive, third party, etc.	Dog Breeders Inc. Co. Ltd., Beacon House, Landsdown, Bournemouth, Dorset
Information about rare breeds	Rare Breeds Survival Trust, National Agricultural Centre, Stoneleigh Park, Warwickshire CU8 2LZ
Goat rugs and coats	Gillrugs, Old Cider Lodge, Kilmington, Axminster, Devon
Tanning of hides for rugs, etc.	Wild Life Tanners Ltd., Tel: 0161 775 9498
Veterinary herbalist	Dorwest Herbs, Veterinary Section, Shipton Gorge, Bridport, Dorset
Papers, conferences, advice, etc on cheese making	Cheese Makers Association, PO Box 256A, Thames Ditton, Surrey KT7 ORH
Animal Homeopathy information	Homeopathy Society for Animal Welfare, Llanrhidian, Gower, West Glamorgan SA3 1HA
Better known suppliers of homeopathic products	Weleda (UK) Ltd., Heanor Road, Ilkeston, Derbyshire DE7 8DR
	Ainsworth, 38 Cavendish Street, London W1 7LH
	A. Nelson & Co. Ltd., 5 Endeavour Way, Wimbledon, London SW19 1UH
NKF AI and breeding services	Mr L. F. Jenner and Mr R. T. Pepper, Nut Knowle Farm, Worlds End, Gun Hill, Horam, East Sussex TN 21 0LA

APPENDIX 2

Affiliated Societies of the British Goat Society 1994

ANGLO NUBIAN BREED SOCIETY
Mrs A. Carrier, No. 2 Bowling Green Cottages, Old Worcester Road, Albrighton, Wolverhampton.

AVON COUNTY GOAT CLUB
Miss J. Macleod, Chestnut Tree Cottage, Watery Lane, Doynton, Bristol.

AYRSHIRE GOAT CLUB
Miss J. Cain, c/o West Burnhead, Galston, Ayrshire.

BATH & DISTRICT GOATKEEPERS SOCIETY
Mrs H. Tizzard, Folly Farm, Cold Ashton, Chippenham, Wiltshire.

BEDS & HERTS GOAT SOCIETY
Mrs D. Padian, Bury Leys Farm, London Lane, Houghton Conquest, Bedfordshire.

BERKS & OXON GOATKEEPERS FEDERATION
Mrs J. Holgate, Greystones, New Inn Rd., Beckley, Oxford.

BINGLEY DAIRY GOAT SOCIETY
Mrs A. Speight, 6 New Road, Denholme, Nr Bradford, West Yorkshire.

BRECKLAND DISTRICT GOATKEEPERS SOCIETY
Mrs D. Carter, Orchard House, Massingham Road, Weasenham, All Saints, King's Lynn, Norfolk.

BRITISH ALPINE BREED SOCIETY
Mrs Head, The Old Tanyard, Pound Hill, Corsham, Wiltshire.

BRITISH CASHMERE SOCIETY
Mrs D. Marsh, 23 Damhead, Lothianburn, Nr Edinburgh, Scotland.

BRITISH SAANEN BREED SOCIETY
Mrs N. Tye, Westways, Greatfield, Wooton Bassett, Wilts.

BRITISH TOGGENBURG SOCIETY
Mrs L. Newton, Dogtree Bank Farm, Grosmont, Whitby, North Yorkshire.

BUCKINGHAMSHIRE GOAT SOCIETY
Mrs C. Turner, Meadacre Cottage, Kimblewick, Aylesbury, Buckinghamshire.

BUCKLEBURY MILK RECORDING GROUP
Mrs P. Dalton, Brambles, Green Lane, Pamber Green, Nr Basingstoke, RG26 6AD.

BURY ST EDMUNDS & DISTRICT GOAT SOCIETY
Mrs S. Mitcham, 5 Harpers Estate, Nayland, Nr Colchester, Essex.

CAMBS. & DISTRICT GOAT SOCIETY
Mrs J. Collin, French Hall Bungalow, Moulton, Newmarket, Suffolk.

CAPRINE OVINE BREEDING SERVICES
Mr M. Clabburn, Somerset Cattle Breeding Centre, Ilminster, Somerset.

CHESHIRE DAIRY GOAT SOCIETY
Mrs D. Rudkin, 1 Dingle Hollow, Compstall Road, Romiley, Stockport, Cheshire.

CHILTERNS GOAT SOCIETY
Mrs N. Cripps, c/o Fleur De Lys, Dagnall, Berkhamstead, Herts.

CLEVELAND DAIRY GOAT SOCIETY
Mrs L. Weedy, 23 Glaidale road, Yarm-On-Tees, Cleveland.

COLCHESTER, SUDBURY & DISTRICT GOAT CLUB
Miss P. Minter, Avon, Park Lane, Langham, Colchester, Essex.

COLERAINE & DISTRICT GOAT CLUB
Mr W. Moody, 119 Cashel Road, Macosquin, Co. Londonderry, N.I.

CONWY & DISTRICT GOAT CLUB
Mrs J. Taylor, Warrandyte Cottage, Conwy Old Road, Dwygyfylchi, Penmaenmawr, Gwynedd.

CORNWALL GOATKEEPERS ASSOCIATION
Mrs S. Richardson, Wood Mill, Prideaux Road, St Blazey, Par, Cornwall.

DARTMOOR MILK RECORDING CLUB
Mrs G. Doe, West Week Farm, Chulmleigh, Devon.

DERBYSHIRE GOAT CLUB
Mrs C. Woodward, Headhouse Farm, Mapperley, Derbyshire.

DEVON GOAT SOCIETY
Mrs S. Turgoose, Butterford Mill House, Diptford, Totnes, Devon.

DORSET GOAT CLUB
Ms L. Townsend, Cosmore House, Cosmore, Dorchester, Dorset.

DUKERIES GOAT SOCIETY
Mrs A. Woodward, 11 Dendy Drive, Woodbeck, Retford,
Nottinghamshire.

DUMFRIES & GALLOWAY G.K. ASSOCIATION
Miss K. Wakefield-Richmond, St Michaels, Lockerbie, Dumfriesshire.

DURHAM DAIRY GOAT SOCIETY
Mrs J. Harbour, The Bungalow, Byers Green, Spennymoor, Co. Durham.

FENLAND GOATKEEPERS ASSOCIATION
Mr D. Rayner, Butt House, Second Marsh Road, Walsoken, Wisbech,
Cambridgeshire.

FROME & DISTRICT GOATKEEPERS CLUB
Mrs M. Garton, 83 Water Lane, Horningsham, Warminster, Wiltshire.

GLAMORGAN GOAT CLUB
Mrs C. Hopkin, 56A Plymouth Road, Penarth, South Glamorgan, Wales.

GLOUCESTERSHIRE GOAT SOCIETY
Miss R. Candy, Rose Cottage, Waterley Bottom, North Nibley,
Gloucestershire.

GOLDEN GUERNSEY GOAT SOCIETY
Mrs P. Dalton, Brambles, Green Lane, Pamber Green, Basingstoke,
Hampshire.

GRAMPIAN GOAT CLUB
Mrs S. Cormack, Langley, Edingight Grange, Keith, Banffshire.

GUERNSEY GOAT SOCIETY
Mrs P. Degaris, The Maylands, Les Francais, Vale, Guernsey, C.I.

HAMBLETON DAIRY GOAT SOCIETY
Mrs M. Holmes, Corn Cob, Thorton-le-Moor, Northallerton, North
Yorkshire.

HAMPSHIRE GOAT CLUB
Mrs L. Dewstow-Newitt, Cornerways, Norleywood, Lymington,
Hampshire.

HARNESS GOAT SOCIETY
Mrs A. Cox, Poplar Ridge, Cornwall's Hill, Lambley, Nottinghamshire.

HEREFORDSHIRE GOAT CLUB
Mrs M. Morgan, Sunnybank, Clyro, Hereford.

HIGHLAND GOAT CLUB
Mrs D. Wright, Boxer's Croft, Abriachan, Inverness-Shire.

HULL & EAST RIDING GOAT SOCIETY
Mrs A. Musgreave, Justmistim, Remington Avenue, Catfoss, Seaton, Hull, East Yorkshire.

IPSWICH & DISTRICT GOAT CLUB
Mrs J. Cox, Hawkdene, Hadleigh Road, Elmsett, Ipswich, Suffolk.

IRISH GOAT CLUB
Mrs M. Timpson, Turnings, Straffan, County Kildare, Eire.

IRISH GOAT PRODUCERS ASSOCIATION
Mrs H. Malone, Kilbride, County Carlow, Eire.

ISLE OF WIGHT GOAT CLUB
Miss P. Young, Burna-By, 13 Quay Lane, Brading, Sandown, Isle of Wight.

JERSEY GOAT SOCIETY
Mrs S. Robins, Charleston, Rue Rouge Cul, St Laurence, Jersey, C.I.

KENT GOAT CLUB
Mrs P. Fogden, Aristocats Cattery, Wey Street, Snave, Ashford, Kent.

LANCASHIRE DAIRY GOAT SOCIETY
Mrs L. Fyles, Quarry House, Southport Road, Scarrisbrick, Ormskirk, Lancashire.

LEADER MILK RECORDING CLUB
Mr J. O'Reilly, Alasty, Straffan, Co. Kildare, Eire.

LINCOLNSHIRE GOAT SOCIETY
Mrs H. Benson, Hilltop, Newton by Toft, Market Rasen, Lincolnshire.

MID ESSEX GOAT CLUB
Mrs M. Butterfield, Woodbine Cottage, Honey Lane, Waltham Abbey, Essex.

MIDSHIRES GOAT CLUB
Mrs L. Doyle, 2 Ascote Cottages, Chapel Ascote, Ladbroke, Leamington Spa, Warwickshire.

NORFOLK & SUFFOLK GOAT SOCIETY
Mrs P. Leggett, Perry Childs Farm, Hall Road, Panfield, Nr Braintree, Essex.

NORTH LEICESTERSHIRE GOAT CLUB
Mrs P. Spilsbury, 45 Littleglen Road, Glen Parva, Leicester.

NORTH MIDLAND MILK RECORDING CLUB
Messrs R. & M. Cooper, 2 Newtop Farm, Broomfield Estate, Morley, Derbyshire.

NORTH STAFFORDSHIRE GOAT SOCIETY
Mrs P. Clee, Russells Bank Farm, 78 Upper Way, Upper Longdon, Nr Rugeley, Staffordshire.

NORTH WALES GOAT SOCIETY
Mrs Jill Barlow, Yr Ynys Uchaf, Mynydd, Llandegai, Bangor, Gwynedd.

NORTHERN ENGLAND GOAT CLUB
Mrs G. Laing, Twislehope, Hermitage, Hawick, Roxborough.

NORTHERN IRELAND GOAT CLUB
Mr Hanna, 35 Mullaghdrin Road, Dromora, Dromore, Co. Down, N. Ireland.

NORTHUMBRIAN DAIRY GOAT SOCIETY
Mrs H. Horn, 32 Spenburn, High Spen, Rowlands Gill, Tyne & Wear.

NORWICH & DISTRICT GOAT CLUB
Mrs J. Cunnington, Trunch Rectory, North Walsham, Norfolk.

NOTTINGHAMSHIRE GOAT CLUB
Mrs V. Hardy, 1 Cottage Ashfield School, Sutton Road, Kirkby-in-Ashfield, Notts, NG17 8HR.

ORKNEY GOAT SOCIETY
Mrs I. Rendall, East Nistaben, Stennes, Orkney.

PENINSULA GOAT SOCIETY
Mrs L. Hill, Northwethel, Little Carbarrack, Redruth, Cornwall, TR16 5RS.

PENNINE GOAT CLUB
Mrs O. Fletcher, 6 Hillcrest View, Denholme, Bradford, West Yorkshire.

PONTEFRACT & DISTRICT GOAT CLUB
Mr G. F. Dronfield, The Ramblers, 25 Bar Road South, Beckingham, Doncaster, Yorkshire.

QUINTET MILK RECORDING CLUB
Mr G. George, Ardross Cottage, Checkendon, Reading, Berkshire.

SAANEN BREED SOCIETY
Mrs J. Tomlinson, 117 Lakes Lane, Newport Pagnell, Buckinghamshire.

SCOTTISH GOATKEEPERS FEDERATION
Mrs F. Fairley, Ladeside House, Gallow Hill Road, Kinross, Kinross-Shire.

SELBY MILK RECORDING GROUP
Mrs L. Bugler, Oakfield Kennels, Skiff Lane, Holme-Upon-Spalding Moor, York.

SHROPSHIRE GOATKEEPERS SOCIETY
Mrs J. Mantle, Rowe Farm, Rowe Lane, Stanton Long, Much Wenlock, Shropshire.

SOMERSET DAIRY GOAT CLUB
Mrs A. Lock, Wayside, Catsham, Nr Baltonsborough, Somerset.

SOUTH LEICESTERSHIRE GOAT SOCIETY
S. Walters, Medbourne Lodge Cottage, Medbourne, Market Harborough, Leicestershire.

SOUTH WALES MILK RECORDING GROUP
Mr M. Elphick, Bryn Y Coed, Crwbin, Kidwelly, Dyfed.

SOUTH WEST WALES GOAT CLUB
Mr G. Jones, Trefangor Farm, Llanddewi, Velfrey, Narbeth, Dyfed.

SOUTHERN SCOTLAND GOAT SOCIETY
Mr R. Weir, Blinkbonny Farm, Currie, Mid Lothian.

SURREY GOAT CLUB
Mrs C. Hord, 24a Pilgrims Way, E. Otford, Sevenoaks, Kent.

SUSSEX COUNTY GOAT CLUB
Mrs S. Thompson, Holt Valley Farm, Underhill Lane, Clayton, Hassocks, West Sussex.

SWAFFHAM & DISTRICT GOAT CLUB
Mrs L. Jermy, Half Moon, The Street, Bintree, Dereham, Norfolk.

TOGGENBURG BREEDERS SOCIETY
Mr C. Chislett, Penponds, Tregavethan, Truro, Cornwall.

WAKEFIELD MILK RECORDING GROUP
Mrs P. Thorpe, Rhvddings Farm, Wakefield Road, Ackworth, West Yorkshire.

WARWICKSHIRE GOAT SOCIETY
Mrs L. Tippett, Meadowcroft, High Cross Lane, Shrewley, Warwick.

WAVENEY VALLEY GOAT CLUB
Miss B. Chivers, Ourway, Cooks Lane, Redenhall, Harleston, Norfolk.

WELSH & MARCHES GOAT SOCIETY
Mrs P. Manning, Pentwyn Bach, Newchurch, Chepstow, Gwent.

WESSEX GOAT CLUB
Mrs M. Smart, The Coach House, Rousdon, Lyme Regis, Dorset.

WEST MIDLANDS GOAT SOCIETY
Mrs A. Carrier, 2 Bowling Green Cottages, Old Worcester Road, Albrighton, Wolverhampton.

WILTSHIRE GOAT SOCIETY
Mrs J. Findlay, 58 Pelch Lane, Seend Cleeve, Melksham, Wiltshire.

WOOLMER MILK RECORDING CLUB
Mrs M. Metcalfe, Shepherds Cottage, Hollywater, Bordon, Hampshire.

WORCESTERSHIRE GOAT SOCIETY
Mrs M. Johnson, 9B Springbank, Hindlip, Worcs.

WYE VALLLEY GOAT CLUB
Ms D. Courtney-Hart, Tan-y-Lan, Railwayside, Clydach, Nr
Abergavenny, Gwent.

YORKSHIRE GOAT SOCIETY
Mrs O. Fletcher, 6 Hill Crest View, Denholme, Bradford, West Yorkshire.

APPENDIX 3

Further Reading

In compiling this list of books, booklets and leaflets, the intention has been to produce a reasonably comprehensive listing of material which is readily accessible to the interested reader in the UK. Whilst some of the publications given below are not necessarily about goats or goatkeeping, they all contain (explicitly or implicitly) information which is relevant.

The appearance of a publication in this list does not necessarily imply recommendation. The abbreviations (B) or (L) before a title indicate that the publication is a booklet or leaflet respectively.

General
'Code of Recommendations for the Welfare of Goats' (PB 0081), which all goatkeepers are required by law to have a copy of, available free from MAFF Publications, London SE99 7TP, phone 0181-694 8862.
'The Goatkeepers' Veterinary Book' by P. Dunn, Farming Press, 3rd ed, 1994, 210 pages.
'Goat Farming' by A. Mowlem, Farming Press, 1988, 192 pages.
'Goat Husbandry' by D. Mackenzie, revised by R. Goodwin, Faber, 1993, £9.99 + p.&p.
'Goat Production in the Tropics' by C. Devendra and M. Burns, Commonwealth Agricultural Bureaux, 1970.
(B) 'Goatkeeping', BGS, 1976.
(B) 'Breeds of Goats', BGS, 1990, £1.50 + p.&p.
(L) 'Life Story of a Goat', BGS.
'Goats for Fibre', National Angora Stud.
'Goat Breeding & Kid Rearing' with Hilary Matthews, Farming Press video, 1995.
'Semen Collection for AI', Mr L. F. Jenner, Nut Knowle Farm, Worlds End, Gun Hill, Horam, East Sussex TN21 0LA.
List of inseminators etc. and list of males from Caprine and Ovine Breeding Society. List of licensed inseminators also from the British Goat Society.

Dairying
'Making Cheeses' by S. Ogilvy, Batsford, 1976.
'Backyard Dairy Book' by A. Singer and L. Street, Prism Press, 1975, 87 pp.
'Cheesecraft', Rita Ash, BGS, £5.00 + £1.00 p.&p.

Feeding and Management

(L) 'Wild Food for Goats', BGS, 10p + p.&p.

'Electric Fencing' (Bulletin 147), HMSO, 1969, 36 pp.

'British Poisonous Plants' (Bulletin 161), HMSO, 1970, 131 pp, 70p + p.&p.

'Rations for Livestock' (Bulletin 48), HMSO, 1970, 134 pp, 70p + p.&p.

Medical Uses of Goats' Milk

'Goats Milk for the Allergic Child', BGS, 15p + p.&p.

'Cows' Milk Intolerance, an alternative', Douglas & Tricia Lock.

Serial Publications

British Goat Journal, BGS (11 issues per year).

British Goat Society Year Book, BGS (annually).

British Goat Society Herd Book, BGS (annually).

Various newsletters and magazines published by local goatkeepers' societies.

BGS Publications are normally available from Secretaries of local goatkeeping organisations or by post from:

The Secretary, British Goat Society, 34–36 Fore St, Bovey Tracey, Newton Abbott, Devon TQ13 9AD. Telephone: 01626 833 168. The Scientific Officer of the Society will give information on rules and regulations of the EU dealing with goats: Mrs R. Goodwin, Southside Cottage, Brook Hill, Albury, Guildford, Surrey GU5 9JD. Telephone: 01483 202 159.

Copies of HMSO Bulletins are available from local booksellers, HMSO bookshop agents or by post from:

HMSO Publications Centre, PO Box 276, London SW8 5DT.

Up-to-date prices and postal charges, together with details of other publications of interest are given in the free catalogue (Sectional List 1: Government Publications: Agriculture) obtainable from the same address.

Telephone orders: 0171 873 9090

Fax orders: 0171 873 8200

General enquiries: 0171 873 0011

For Ministry of Agriculture, Fisheries and Food (MAFF) publications, the helpline number is 01645 335 777.

INDEX

Farming Press Books & Videos

Below is a sample of the wide range of agricultural and veterinary books and videos we publish. For more information or for a free illustrated catalogue of all our publications please contact:

**Farming Press,
Miller Freeman UK Ltd,
2 Wharfedale Road, Ipswich, Suffolk, IP1 4LG,
United Kingdom.
Tel: (01473) 241122 Fax (01473) 242222
E-mail: farmingpress@dotfarming.com
Website: http://www.dotfarming.com**

The Goatkeeper's Veterinary Book (3rd edition)   **Peter Dunn**
The standard veterinary reference work on diagnosis, treatment and prevention of goat ailments.

Goat Farming   **Alan Mowlem**
Covers all aspects for those considering goats as an alternative enterprise. The book links scientific theory to commercial working practice.

Goats of the World   **Valerie Porter**
Over 520 breeds and types of goat are described in this comprehensive book, a major contribution to the study of goat breeds worldwide.

Poultry at Home (VHS Video)   **Victoria Roberts**
A beginner's guide to poultry health and management.

A Manual of Lambing Techniques **Agnes Winter and Cicely Hill**
Advice is given on all aspects of lambing, including normal and rare presentations, pre- and post-lambing prolapses and possible problems and emergencies in the first few days.

An Introduction to Keeping Sheep (2nd edition) **Jane Upton and
Dennis Soden**
The skills and techniques of caring for sheep for newcomers.

Farming Press is a division of Miller Freeman UK Ltd which provides a wide range of media services in agriculture and allied businesses. Among the magazines published by the group are *Arable Farming, Dairy Farmer, Farmers Guardian, Farming News* and *Pig Farming*. For a specimen copy of any of these please contact the address above.